Split Federated Learning for Secure IoT Applications

Other related titles:

You may also like

- Abdel-Basset | Explainable Artificial Intelligence for Trustworthy Internet of Things | 2024
- PBSE020 | Abd El-Latif | Artificial Intelligence for Biometrics and Cybersecurity | 2023
- PBPC057 | Alazab | AIoT Technologies and Applications for Smart Environments | 2022
- PBSE016 | Tyagi | Machine Learning, Blockchain Technologies and Big Data Analytics for IoTs: Methods, technologies and applications | 2022

We also publish a wide range of books on the following topics:
Computing and Networks
Control, Robotics and Sensors
Electrical Regulations
Electromagnetics and Radar
Energy Engineering
Healthcare Technologies
History and Management of Technology
IET Codes and Guidance
Materials, Circuits and Devices
Model Forms
Nanomaterials and Nanotechnologies
Optics, Photonics and Lasers
Production, Design and Manufacturing
Security
Telecommunications
Transportation

All books are available in print via https://shop.theiet.org or as eBooks via our Digital Library https://digital-library.theiet.org.

IET SECURITY SERIES 025

Split Federated Learning for Secure IoT Applications

Concepts, frameworks, applications and case studies

Edited by
Gururaj Harinahalli Lokesh, Geetabai S. Hukkeri,
N.Z. Jhanjhi and Hong Lin

The Institution of Engineering and Technology

About the IET

This book is published by the Institution of Engineering and Technology (The IET).

We inspire, inform and influence the global engineering community to engineer a better world. As a diverse home across engineering and technology, we share knowledge that helps make better sense of the world, to accelerate innovation and solve the global challenges that matter.

The IET is a not-for-profit organization. The surplus we make from our books is used to support activities and products for the engineering community and promote the positive role of science, engineering and technology in the world. This includes education resources and outreach, scholarships and awards, events and courses, publications, professional development and mentoring, and advocacy to governments.

To discover more about the IET please visit https://www.theiet.org/.

About IET books

The IET publishes books across many engineering and technology disciplines. Our authors and editors offer fresh perspectives from universities and industry. Within our subject areas, we have several book series steered by editorial boards made up of leading subject experts.

We peer review each book at the proposal stage to ensure the quality and relevance of our publications.

Get involved

If you are interested in becoming an author, editor, series advisor, or peer reviewer please visit https://www.theiet.org/publishing/publishing-with-iet-books/ or contact author_support@theiet.org.

Discovering our electronic content

All of our books are available online via the IET's Digital Library. Our Digital Library is the home of technical documents, eBooks, conference publications, real-life case studies and journal articles. To find out more, please visit https://digital-library.theiet.org.

In collaboration with the United Nations and the International Publishers Association, the IET is a Signatory member of the SDG Publishers Compact. The Compact aims to accelerate progress to achieve the Sustainable Development Goals (SDGs) by 2030. Signatories aspire to develop sustainable practices and act as champions of the SDGs during the Decade of Action (2020–30), publishing books and journals that will help inform, develop, and inspire action in that direction.

In line with our sustainable goals, our UK printing partner has FSC accreditation, which is reducing our environmental impact on the planet. We use a print-on-demand model to further reduce our carbon footprint.

Published by The Institution of Engineering and Technology, London, United Kingdom

The Institution of Engineering and Technology (the "**Publisher**") is registered as a Charity in England & Wales (no. 211014) and Scotland (no. SC038698).

First published 2024

The Institution of Engineering and Technology
Futures Place
Kings Way, Stevenage
Herts, SG1 2UA, United Kingdom

www.theiet.org

British Library Cataloguing in Publication Data
A catalogue record for this product is available from the British Library

ISBN 978-1-83953-945-9 (hardback)
ISBN 978-1-83953-946-6 (PDF)

Typeset in India by MPS Limited

Cover image: Cover Image: Olemedia/E+ via Getty Images

Contents

Preface

This book is a descriptive summary of applications and challenges of split federated learning with various case studies from diverse authors across the globe.

The authors of Chapter 1 elaborate on the fundamentals of federated learning (FL) and split learning which are the two common approaches to distributed machine learning. This chapter introduces splitfed learning, a novel approach that combines the two abovementioned approaches while eliminating the resulting drawbacks, as well as a refined structural configuration including distinct security and PixelDP to improve data privacy and durability.

In Chapter 2, the authors have discussed an overview of the cutting-edge technologies for integrating these FL approaches into an edge computing-based IoT setting. Furthermore, it addresses some existing challenges and also discusses future research directions to spark further exploration within the academic community.

The authors of Chapters 3 and 4 provide an overview of blockchain-driven splitfed learning for data protection in IoT setting and splitfed learning methods for natural language processing.

In Chapters 5 and 6, the authors are concerned with the role of splitfed learning in recommendation systems. Delving into the applications of RIS within the context of 6G, various configurations of RIS-assisted wireless systems are scrutinized, covering diverse scenarios, system and fading models, as well as performance metrics and objectives, in a comprehensive and methodical manner.

The authors of Chapters 7 and 8 have tried to enhance computational performance in healthcare through federated learning approach and splitfed learning for multimodal emotion detection. Understanding emotions is essential for positive social interactions. It also helps to better gauge how customers feel about a product and understand how their behaviours are changing. For this reason, the market for emotion recognition is currently booming, despite having a lot of gaps in data.

The authors of Chapter 9 provide an up-to-date evaluation of predictive models used in forecasting student performance and learning outcomes in the educational sector. It examines modern machine learning and deep learning models for predicting college dropouts, utilizing cost-effective large-scale administrative data as an early warning system. In Chapter 10, the authors have focused on exploring the application of splitfed learning in the realm of smart transportation, focusing on its ability to enhance the privacy and efficiency of automated transportation systems.

In Chapters 11 and 12, the authors have discussed the case study of splitfed learning for smart grids and smart agriculture. Splitfed learning enables immediate improvements through localized model updates, promotes data privacy by retaining control over sensitive information within individual components and enhances the overall efficiency and resilience by leveraging diverse data sources.

Gururaj H.L.

About the editors

Gururaj Harinahalli Lokesh is an associate professor in the Department of Information Technology and the founder of the Wireless Internetworking Group (WiNG) at the Manipal Institute of Technology, Bengaluru, India. His research interests include blockchain technology, cyber security, wireless sensor networks, ad-hoc networks, IoT, data mining, cloud computing and machine learning. He is an editorial board member of the *International Journal of Blockchains and Cryptocurrencies*. He has published over 110 research papers in international journals and presented 20 papers at international conferences. He is a senior member of the IEEE, lifetime member of ISTE and CSI, a professional member of the ACM and an ACM distinguished speaker. He holds a PhD degree in Computer Science and Engineering from Visvesvaraya Technological University, Belagavi, India.

Geetabai S. Hukkeri is an assistant professor in the Department of Computer Science and Engineering at the Manipal Institute of Technology, Bengaluru, India. Her research interests include artificial intelligence, deep learning, machine learning, big data, computer vision and multimedia information retrieval. She is a member of ACM. She holds a PhD degree in Computer Science and Engineering from Visveswaraya Technological University, Belagavi, India.

N.Z. Jhanjhi is a professor at the School of Computer Science, Taylor's University, Malaysia, and is the program director for Postgraduate Research Degree Programmes in Computer Science. Additionally, he directs the Center for Smart Society (CSS5), playing a crucial role in advancing research and academic excellence at the university. He is a distinguished academician, researcher and scientist in computer science, specializing in cybersecurity.

Globally recognized among the top 2% of research scientists and ranked among the top three computer science researchers in Malaysia, Prof. Jhanjhi was named an Outstanding Faculty Member by MDEC Malaysia in 2022. His extensive research contributions are reflected in his highly indexed publications in prestigious journals, with a collective research impact factor exceeding 1000. He has authored and edited over 50 research books with renowned publishers including Springer, Elsevier, Taylor and Francis, Wiley and IGI Global USA.

Hong Lin is a professor of Computer Science at the University of Houston, Downtown, TX, USA, where he has established the Grid Computing Lab through an NSF MRI grant. He previously worked on large-scale computational biology at

Purdue University and network security at Nokia, Inc. His areas of research include parallel computing, multi-agent systems and affective computing. He has published over 100 research papers in international journals and conferences. He is a senior member of the ACM. He holds a PhD in Computer Science from Purdue University, USA.

Chapter 1

Introduction to federated learning, split learning and splitfed learning

Geetabai S. Hukkeri[1], *Gururaj H.L.*[2], *N.Z. Jhanjhi*[3] *and Hong Lin*[4]

Federated learning (FL) and split learning (SL) are two prevalent methods of distributed machine learning (ML). In both approaches, users train and test ML models without sharing raw data. However, SL offers enhanced model privacy compared to FL due to its separation of the ML model architecture between clients and the server. Furthermore, because of the split model, SL is a superior choice for resource-constrained applications. However, because of the relay-based training across numerous clients, SL performs slower than FL. In this regard, this chapter introduces splitfed learning, a new technique that incorporates the two approaches while eliminating the resulting drawbacks, as well as a refined structural configuration including distinct security and PixelDP to improve data privacy and model its durability. Splitfed is useful in resource-constrained contexts when complete model training and installation are impossible and quick model training time is necessary to continually update the global model on an evolving dataset. These contexts define different domains.

Keywords: Internet of Things; Federated learning; Split learning; Machine learning; Splitfed learning

1.1 Introduction

The rapid development of networking as well as distributed computing capabilities has empowered the widespread mobilisation of the Internet of Things (IoTs), in which a massive amount of data created by various consumer devices is analysed

[1]Department of Computer Science and Engineering, Manipal Institute of Technology Bengaluru, Manipal Academy of Higher Education, India
[2]Department of Information Technology, Manipal Institute of Technology Bengaluru, Manipal Academy of Higher Education, India
[3]School of Computer Science (SCS), Taylor's University, Malaysia
[4]Department of Computer and Mathematical Sciences, University of Houston-Downtown, Houston, TX, USA

via multiple applications. As datasets grow higher, training machine learning (ML) algorithms increasingly necessitates sharing model parameter optimisation across numerous processors.

'Distributed Collaborative Machine Learning' (DCML) is notable in view of its inherent data privacy advantages [1]. In contrast to traditional methods that centralise and access data, DCML allows ML without needing data transfer from data custodians to untrusted parties. In this approach, analysts do not have access to the source data; rather, the ML framework is sent to the data custodian for computing. Additionally, DCML supports processing across multiple platforms or servers, including distributed devices.

Federated learning (FL) [2,3] and split learning (SL) [4] are among the most widely adopted DCML techniques. FL involves training an ML model on distributed clients using their local data, followed by aggregating these locally trained models on a server to create a global model. A key advantage of FL is its ability to facilitate parallel and thus efficient training of ML models across numerous users. Nonetheless, FL necessitates clients to run the entire ML model, which can be prohibitively expensive for resource-constrained devices such as those in the IoT. This challenge is particularly pronounced with deep learning models. Additionally, during training, both the server and clients have full access to local and global models, posing a privacy risk from the model's point of view.

SL was developed to address these concerns. SL partitions the entire ML model into smaller segments, which are then trained separately on a server and on distributed clients using local data. This approach significantly minimises the computing burden over resource-constrained devices by allowing them to train only a portion of the network, unlike in FL where the entire model must be executed [5]. Additionally, in SL, clients do not have access to the centralised model and vice versa, enhancing privacy.

Despite the benefits of SL, there is a major drawback. Since only one client communicates with the server at one time in relay-based training with SL, this leads to idle resources on the clients' side, resulting in a considerable increase in training overhead with numerous clients. This chapter offers a unique architecture termed splitfed learning (SFL) to overcome these difficulties in FL and SL. SFL explores the benefits of FL and SL at the time of emphasising data privacy and model strength.

FL [6] is an ML method that allows model training to take place across several decentralised edge devices like mobile phones, IoT devices without the need for raw data to be exchanged. Traditional ML algorithms necessitate centralising all data in one area, which can be inconvenient for huge datasets, as well as inefficient due to excessive communication costs. In place of transmitting source data to a centralised server, FL sends the model to each device. On the basis of its local data, each device computes an update to the model. These modifications are subsequently sent back to the central server, that collects them to update the centralised model. This procedure is performed iteratively, allowing the model to benefit from the pooled understanding of all participating machines while keeping data localised and private. FL has numerous applications, including finance, healthcare and IoT, where data confidentiality and communication effectiveness are critical.

SL [4] is a strategy that falls under the umbrella of FL. SL splits the model into two components: a local half and a centralised part. The local component processes the raw data on the user's device and only transmits the intermediate approximations to the server-side component. The server-side component merges these preliminary representations from different systems and maintains the global model. This framework guarantees that the raw data is never sent outside of the local device, protecting user privacy. SL is especially beneficial when the raw data is confidential or too vast to be efficiently delivered over the network. SL lowers communication bandwidth utilisation and privacy concerns by keeping data processing local and communicating only essential information.

FL enables simultaneous ML model training across multiple clients, whereas SL offers improved privacy by separating the ML model across clients and the server. This separation provides two key advantages. Primarily, it enhances privacy of the model since clients do not have access to the centralised model and vice versa.

Moreover, training a specific section of the network locally reduces computational demands compared to running the entire network, which is essential for ML operations on devices with limited resources. However, SL tends to be slower than FL because of the sequential process of training ML models across clients. Upon weighing the strengths and weaknesses of FL and SL, a pertinent question emerges: why not integrate both SL and FL to leverage their respective advantages?

As a response, SFL, a revolutionary architecture that mixes FL and SL, has been presented. SFL is a good option that gives greater security than FL and is quicker than SL with comparable model accuracy and effectiveness in communication. Furthermore, SFL offers distributed processing among clients with the same modest computational power as SL, but with a five-fold increase in training speed, according to empirical research.

Our contribution in this work includes a detailed study of SL and FL, as well as the presentation of an SFL that is favourable concerning performance and privacy compared to the latest advanced resolutions.

1.2 Related works

This section covers the core concepts of technologies associated with FL, SL and SFL. Additionally, recent advancements in this research area are also addressed.

1.2.1 Federated learning

FL is a decentralised ML approach where every client possesses the same neural network and training data. During each training round, every client conducts forward and backward passes on its model to compute revised parameters. These updated values are then sent to a central server, which combines them with updates from other clients using a method like FedAvg [6] to generate an overall update.

The overall update is essentially the sum of all distinct client weights, which is then distributed back to each client to initiate the next training phase with uniform values. This iterative process continues till the model meets the requirements. FL ensures model training without necessitating the transmission of local training data between clients and the server.

Despite its advantages, FL has been vulnerable to security issues such as model poisoning attacks that significantly impact model accuracy [7–9]. Notably, the Little is Enough (LIE) Attack [10] leverages mean and standard deviation of updates to introduce harmful code. Virat Shejwalkar [7] has developed a dynamic optimisation-based approach that is more potent than previous methods.

Recent studies have also explored combining deep learning or FL with edge computing [11–14]. Privacy protection is a critical topic within FL [15–19], discussed extensively in [20–24]. Li *et al.* [20] addressed privacy concerns during model sharing by proposing sketch algorithms for sketching-based FL that balances privacy and accuracy.

To further address privacy, authors in [21] introduced a privacy-enhanced FL system that contributes to efficient and privacy-preserving training. In [22], a system combining procreative oppositional networks with a multitask identifier is proposed to counter user-level privacy leaks against malicious server attacks in FL. Nasr *et al.* [23] conducted thorough privacy assessment using white-box inference attacks. The authors in [24] applied FL principles to recommender systems with matrix factorisation, ensuring system functionality and anonymity.

Furthermore, as fog computing and edge computing have grown in popularity, there have been numerous research concentrating on privacy-preserving crowdsourcing [25,26] and using edge computing or fog computing to increase performance [9,27–31]. The authors of [11] suggested a privacy-preserving deep learning architecture that uses a two-level protection method to safeguard data privacy, including 'Random Projection and Differentially Private' stochastic gradient descent (SGD). In [12], authors created a combined training technique to safeguard the privacy of features. Feature extraction is performed locally in devices such as mobile phones, whereas categorisation is performed in the cloud service. In [13,14], authors suggested control techniques to address the issue of limited resources in IoT devices participating in FL.

1.2.2 Split learning

SL was initially proposed as a solution to address limitations in a client's hardware resources [4,5]. The concept involves dividing a DNN model into two sub-models with specific layer allocations. One sub-model, the client-side model, contains the initial layers and is used by clients to train their local data. The other sub-model, the centralised model, includes the remaining layers and is responsible for aggregating updates from clients.

During each iteration, clients train their local model using their data and send only the necessary activation values (termed 'shredded data') to the server. The server, which possesses the full model but only processes the shredded data, computes the gradient updates for the global model. This allows client data to remain private as the raw data is not transmitted; only processed information is shared.

This approach eliminates the need for each client to train the complete model locally, thereby reducing the training burden on client devices. In traditional SL, however, training happens sequentially, which can lead to increased training time when more clients are involved [32–34].

Jeon *et al.* [35] introduced a parallel method where all clients can contribute simultaneously to the training process, aiming to reduce overall training time. Nonetheless, as the count of clients increases, the training time can still extend due to the sequential nature of centralised model updates, resembling traditional SL in the worst-case scenario.

Zhang *et al.* [36] used the parallel training technique as well as the feature connect method to train the centralised model. When the centralised training process is complete, the gradient is split based on the defeated data size of each client. To change the local model, the relevant gradient is returned to the client. However, in prior studies, each client was assigned the identical model layers, resulting in a waste of processing power for the client with outstanding evaluating capabilities.

1.2.3 Split-federated learning

SFL [37] represents an innovative distributed collaborative ML approach that integrates elements from existing technologies discussed earlier. This method combines the pros of both paradigms while mitigating their cons. Alike to SL, the model is among clients and a server, enabling computations on devices with limited resources. Additionally, similar to FL, an extra server called the Fed Server collects model updates from each client, which are evaluated in parallel during each training process, to update the global client component of the model. The SFL framework is especially well-suited for edge deployments. Extensive research has focused on FL security, particularly regarding model poisoning attacks that alter model gradients on compromised devices to reduce accuracy [7,8,10,38]. This highlights the importance of robust security measures in distributed ML frameworks like SFL. The bigger the dimensionality of a model, the more susceptible it is to an attack because the attacker has more room to disturb the model invisibly. The authors of [39] used poisoning assaults to investigate the constraint of dimensionality in FL. However, research on SL privacy has been minimal. The authors of [40] employed inference-based attacks [41] to retrieve private training data via the client or the server.

In [42], authors introduced SFL to power the advantages of both FL and SL in training split models, aiming to minimise training time and computational demands on client devices. SFL comprises two training processes known as SFLv1 and SFLv2 [42].

Several studies have proposed enhancements to improve training communication efficiency between clients and servers based on SFL methodologies. He *et al.* [43] utilised a similar architecture where the local model handled shallower layers (like ResNet [44]), while the centralised model managed deeper layers. Through knowledge distillation, both local and centralised models learn from each other's predictions. However, sequential training of the centralised model may extend the training duration when many clients are involved.

Pal *et al.* [32] updated the SFL server-side model with shattered data from all clients, while the local model was updated with data from individual clients. An issue of updating imbalance arose when using the same learning rate for both models, suggesting that the centralised model's learning rate should adapt based on the number of clients. The authors of [45] emphasised the necessity in conventional SFL architectures for local models to wait until the server computes gradients of the shattered data before updating.

1.3 Splitfed learning: combining federated learning and split learning

SFL leverages the pros of both SL and FL. SFL integrates FL's strengths, such as parallel processing across distributed clients, with SL's approach of network splitting while training [46–48]. The SFL model is illustrated in Figure 1.1. The diagram on the right illustrates how network training is conducted by the clients and the main server.

In contrast to SL, all clients simultaneously accomplish forward propagation on their local model and then submit the crushed data to the main server. The server, assumed to have sufficient processing resources, then performs forward and back propagation on its server-side algorithm, processing each client's crushed data in a slightly parallel manner.

The processed gradients of the shattered data (i.e., activations) are subsequently transmitted back to the relevant clients for back-propagation. The server updates its model by averaging the gradients obtained from back-propagation on each client's processed data. Once the clients receive the gradient updates, they perform back-propagation on their local models and update their parameters accordingly. Clients then send these updated gradients to the fed server, which

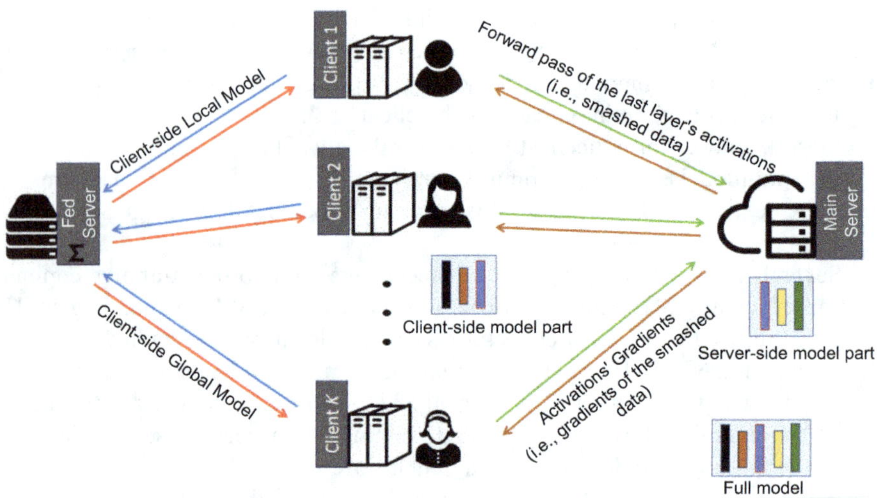

Figure 1.1 Splitfed system model [37]

Table 1.1 Comparative overview of SL, FL and SFL

Learning methods	Client-side training	Model privacy advantage by split model	Model aggregation	Access to raw data	Distributed computing
SL	Sequential	✓	×	×	✓
FL	Parallel	×	✓	×	✓
SFL	Parallel	✓	✓	×	✓

aggregates them using a federated median approach and distributes the results back to all participating clients. This synchronisation of the client-side global model occurs in each iteration of network training. The evaluations performed by the fed server are efficient and are confined within local edge boundaries. For a comparative overview of SL, FL and SFL, refer to Table 1.1.

1.3.1 Why splitfed learning

SFL is an ML technique that tackles privacy concerns by allowing model training without the need for source data sharing. It is especially important in situations where security and confidentiality of data are critical such as healthcare, banking and other businesses with sensitive information. Here's why SFL is employed and what the main benefits are:

- *Data remains local:* The training process is divided into two parts: the model is sent to devices that are local, and the training procedure is performed on the local data without forwarding source data to a central server.
- *No centralised data storage:* The raw data is never stored anywhere other than the local devices, thereby helping to prevent privacy violations and unauthorised access.
- *Lowered attack surface:* Because raw data is not centralised, the chance of an information leak or attack on a centralised server is lowered dramatically.
- *Secure communication:* Typically, communication between local machines and the centralised server is encrypted, offering an additional layer of protection.
- *Reduced data transfer:* Rather than sending vast volumes of unprocessed information to a centralised server for model training, simply model updates (weights) are sent, minimising the quantity of data sent over the network.
- *Data protection law compliance:* SFL learning can assist organisations in complying with data protection regulations and legislation by minimising the exposure of sensitive information.
- *Efficient resource utilisation:* SL enables distributed model training by utilising local compute resources that can be more effective in terms of resource utilisation.
- Federated strategy: SFL learning is sometimes applied in the framework of a FL strategy, in which many parties cooperate to train a model without

disclosing raw data. This is useful when data from several sources is required for a thorough model.

1.3.2 Applications of splitfed learning

Because of its capacity to protect data privacy, reduce transmission bandwidth and permit collaborative model training, the SFL technique offers a wide range of applications. Here are several SFL applications:

1. **Healthcare:**
   - *Health record analysis:* Hospitals and healthcare providers can work together to develop training models for analysing electronic health records without directly disclosing sensitive patient data. This aids in disease prediction, therapy optimisation and overall healthcare service improvement.
   - *Drug development:* Pharmaceutical organisations can employ SFL to construct ML models for drug development without directly sharing confidential chemical or biological data. Collaboration can take place between research institutions and businesses.
2. **Finance:**
   - *Fraud detection:* Banks and financial institutions can collaboratively train fraud detection models without sharing individual transaction details. Each institution can train the model on its local data, ensuring privacy and then share updates to improve fraud detection collectively.
   - *Algorithmic trading:* SFL can be applied to train trading algorithms collaboratively, allowing different trading firms to work together without revealing their specific trading strategies.
3. **Smart cities:**
   - *Traffic management:* Cities can employ SFL to analyse traffic patterns and optimise traffic management systems without centralising sensitive location data. This can lead to more efficient traffic flow and reduced congestion.
   - *Energy Optimisation:* Utilities can use SFL to optimise energy distribution and consumption patterns in smart grids. Local data from different regions can be used collaboratively to enhance energy efficiency.
4. **IoT devices:**
   - *Predictive maintenance:* IoT devices in industries such as manufacturing and transportation can use SFL to predict maintenance needs. Data from various devices can be used to create predictive maintenance models without sending raw data to a central server, ensuring data privacy.
   - *Home automation:* Smart home devices can utilise SFL to improve user experience and energy efficiency. Devices can collectively learn user preferences without sharing individual data, enhancing automation and personalisation.

5. **Personalised learning:** Educational institutions can collaborate to create personalised learning experiences for students. SFL enables the development of models that adapt to individual learning styles without exposing sensitive student data.
6. **Customer behaviour analysis:** Retailers can use SFL to analyse customer behaviour across different locations without sharing specific customer data. This information can be valuable for optimising product placements and marketing strategies.
7. **Virtual assistants:** SFL can be applied to enhance virtual assistant technologies. Instead of sending voice recordings to a central server, individual devices can collaborate to improve speech recognition and language understanding while preserving user privacy.

1.4 Conclusion

FL and SL are two promising cooperative learning technologies that can be combined to enable ubiquitous intelligence in IoT. In this chapter, we first discussed an overview of SL, FL and SFL. We then discuss each of these methods in detail, highlighting their advantages and disadvantages. Furthermore, we review cutting-edge solutions for integrating SL with FL to enhance pervasive intelligence in IoT. The survey highlights substantial advancements across different dimensions of integrating SL and FL, encompassing developments in SFL frameworks, model decomposition methods, enhancements in performance and efficiency, and advancements in privacy safeguards. However, research in this field is still in its early stages, with many unresolved problems requiring further investigation.

References

[1] P. Kairouz, H.B. McMahan, B. Avent, *et al.*, Advances and open problems in federated learning. arXiv:1912.04977, 2019, http://arxiv.org/abs/1912.04977.

[2] J. Konečný, B. McMahan, and D. Ramage, Federated optimization: Distributed optimization beyond the datacenter. arXiv:1511.03575, 2015, https://doi.org/10.48550/arXiv.1511.03575.

[3] H.B. McMahan, E. Moore, D. Ramage, S. Hampson, and B. Agüera y Arcas, "Communication-efficient learning of deep networks from decentralized data." In *Proceedings of the 20th International Conference on Artificial Intelligence and Statistics (AISTATS)*, 2017.

[4] O. Gupta and R. Raskar, "Distributed learning of deep neural network over multiple agents," *Journal of Network and Computer Applications*, 116, 1–8, 2018, https://doi.org/10.1016/j.jnca.2018.05.003.

[5] P. Vepakomma, O. Gupta, T. Swedish, and R. Raskar, "Split learning for health: Distributed deep learning without sharing raw patient data," arXiv:1812.00564, 2018, https://doi.org/10.48550/arXiv.1812.00564.

[6] B. McMahan, E. Moore, D. Ramage, S. Hampson, and B. A. Arcas, "Communication-efficient learning of deep networks from decentralized data." In

Proceeding of the 20th International Conference on Artificial Intelligence and Statistics, PMLR, 1273–1282, 2017.

[7] V. Shejwalkar, A. Houmansadr, P. Kairouz, and D. Ramage, "Back to the drawing board: A critical evaluation of poisoning attacks on production federated learning." In *2022 IEEE Symposium on Security and Privacy (SP)*, IEEE, 1354–1371, 2022.

[8] M. Fang, X. Cao, J. Jia, and N. Gong, "Local model poisoning attacks to byzantine-robust federated learning." In *29th USENIX Security Symposium (USENIX Security 20)*, USENIX Association, 1605–1622, 2020, https://www.usenix.org/conference/usenixsecurity20/presentation/fang.

[9] P. Zhao, H. Huang, X. Zhao, and D. Huang, "P3: Privacy-preserving scheme against poisoning attacks in mobile-edge computing," *IEEE Transactions on Computational Social Systems*, 7(3), 818–826, 2020, doi:10.1109/TCSS.2019.2960824.

[10] G. Baruch, M. Baruch, and Y. Goldberg, "A little is enough: Circumventing defenses for distributed learning." *Advances in Neural Information Processing Systems*, 32, 8635–8645, 2019.

[11] L. Lyu, J. C. Bezdek, X. He, and J. Jin, "Fog-embedded deep learning for the Internet of Things," *IEEE Transactions on Industrial Informatics*, 15, 4206–4215, 2019.

[12] L. Jiang, X. Lou, R. Tan, and J. Zhao, "Differentially private collaborative learning for the IoT edge." In *International Workshop on Crowd Intelligence for Smart Cities: Technology and Applications (CISC)*, 2018.

[13] S. Wang, T. Tuor, T. Salonidis, *et al.*, "Adaptive federated learning in resource constrained edge computing systems," *IEEE Journal on Selected Areas in Communications*, 37(6), 1205–1221, 2019.

[14] X. Wang, Y. Han, C. Wang, Q. Zhao, X. Chen, and M. Chen, "In-edge ai: Intelligentizing mobile edge computing, caching and communication by federated learning," *IEEE Network*, 33(5), 156–165, 2019.

[15] Q. Yang, Y. Liu, T. Chen, and Y. Tong, "Federated machine learning: Concept and applications," *ACM Transactions on Intelligent Systems and Technology (TIST)*, 10(2), 1–19, 2019.

[16] W. Y. B. Lim, N. C. Luong, D. T. Hoang, *et al.*, "Federated learning in mobile edge networks: A comprehensive survey," *IEEE Communications Surveys & Tutorials*, 22, 2031–2063, 2020.

[17] T. Li, A. K. Sahu, A. Talwalkar, and V. Smith, "Federated learning: Challenges, methods, and future directions," *IEEE Signal Processing Magazine*, 37(3), 50–60, 2020.

[18] T. Nishio and R. Yonetani, "Client selection for federated learning with heterogeneous resources in mobile edge." In *ICC 2019-2019 IEEE International Conference on Communications (ICC)*, IEEE, 2019, pp. 1–7.

[19] L. Lyu, H. Yu, and Q. Yang, "Threats to federated learning: A survey," arXiv:2003.02133, 2020, https://doi.org/10.48550/arXiv.2003.02133.

[20] Z. Liu, T. Li, V. Smith, and V. Sekar, "Enhancing the privacy of federated learning with sketching," arXiv:1911.01812, 2019, https://doi.org/10.48550/arXiv.1911.01812.

[21] M. Hao, H. Li, X. Luo, G. Xu, H. Yang, and S. Liu, "Efficient and privacy-enhanced federated learning for industrial artificial intelligence," *IEEE Transactions on Industrial Informatics*,16(10), 6532–6542, 2020.

[22] Z. Wang, M. Song, Z. Zhang, Y. Song, Q. Wang, and H. Qi, "Beyond inferring class representatives: User-level privacy leakage from federated learning." In *IEEE INFOCOM 2019-IEEE Conference on Computer Communications, IEEE*, 2019, 2512–2520.

[23] M. Nasr, R. Shokri, and A. Houmansadr, "Comprehensive privacy analysis of deep learning: Passive and active white-box inference attacks against centralized and federated learning." In *2019 IEEE Symposium on Security and Privacy (SP)*, IEEE, 2019, 739–753.

[24] K. Dolui, I. C. Gyllensten, D. Lowet, S. Michiels, H. Hallez, and D. Hughes, "Towards privacy-preserving mobile applications with federated learning: The case of matrix factorization (poster)." In *The 17th Annual International Conference on Mobile Systems, Applications, and Services*, Seoul, Korea, 624–625, 2019.

[25] F.-J. Wu and T. Luo, "CrowdPrivacy: Publish more useful data with less privacy exposure in crowdsourced location-based services," *ACM Transactions on Privacy and Security (TOPS)*, 23(1), 1–25, 2020.

[26] T. Liang, "Enabling privacy preservation and decentralization for attribute-based task assignment in crowdsourcing," *Journal of Computer and Communications*, 8(4), 81–100, 2020.

[27] Y. He, J. Ni, B. Niu, F. Li, and X. S. Shen, "Privbus: A privacy-enhanced crowdsourced bus service via fog computing," *Journal of Parallel and Distributed Computing*, 135, 156–168, 2020.

[28] J. Zhang, Q. Zhang, and S. Ji, "A fog-assisted privacy-preserving task allocation in crowdsourcing," *IEEE Internet of Things Journal*, 7(5), 8331–8342, 2020.

[29] J. Xu, S. Wang, B. K. Bhargava, and F. Yang, "A blockchain-enabled trustless crowd-intelligence ecosystem on mobile edge computing," *IEEE Transactions on Industrial Informatics*, 15(6), 3538–3547, 2019.

[30] W. Shi, J. Cao, Q. Zhang, Y. Li, and L. Xu, "Edge computing: Vision and challenges," *IEEE Internet of Things Journal*, 3(5), 637–646, 2016.

[31] W. Shi and S. Dustdar, "The promise of edge computing," *Computer*, 49(5), 78–81, 2016.

[32] S. Pal, M. Uniyal, J. Park, *et al.*, "Server-side local gradient averaging and learning rate acceleration for scalable split learning." In *International Workshop on Trustable, Verifiable and Auditable Federated Learning*, Vancouver, BC, Canada, 1–9, 2022.

[33] S. Oh, J. Park, P. Vepakomma, *et al.*, "Locfedmix-sl: Localize, federate, and mix for improved scalability, convergence, and latency in split learning." In *Proceedings of the ACM Web Conference*, Lyon, France, 3347–3357, 2022.

[34] W. Wu, M. Li, K. Qu, *et al.*, "Split learning over wireless networks: Parallel design and resource management," arXiv:2204.08119, 2022, https://doi.org/10.48550/arXiv.2204.08119.

[35] J. Jeon and J. Kim, "Privacy-sensitive parallel split learning." In *International Conference on Information Networking*, Barcelona, Spain, 7–9, 2020.

[36] M. Zhang, L. Qu, P. Singh, J. Kalpathy-Cramer, and D. L. Rubin, "SplitAVG: A heterogeneity-aware federated deep learning method for medical imaging," *IEEE Journal of Biomedical and Health Informatics*, 26(9), 4635–4644, 2022.

[37] C. Thapa, P. C. M. Arachchige, S. Camtepe, and L. Sun, "Splitfed: When federated learning meets split learning." In *Proceedings of the AAAI Conference on Artificial Intelligence*, 36, 8485–8493, 2022.

[38] V. Shejwalkar and A. Houmansadr, "Manipulating the byzantine: Optimizing model poisoning attacks and defenses for federated learning." In *NDSS*, 21–25, 2021.

[39] H. Chang, V. Shejwalkar, R. Shokri, and A. Houmansadr, "Cronus: Robust and heterogeneous collaborative learning with black-box knowledge transfer." 2019, arXiv:1912.11279, https://doi.org/10.48550/arXiv.1912.11279.

[40] D. Pasquini, G. Ateniese, and M. Bernaschi, "Unleashing the tiger: Inference attacks on split learning. In *Proceedings of the 2021 ACM SIGSAC Conference on Computer and Communications Security*, 2113–2129, 2021.

[41] B. Hitaj, G. Ateniese, and F. Perez-Cruz, "Deep models under the GAN: Information leakage from collaborative deep learning." In *Proceedings of the 2017 ACM SIGSAC Conference on Computer and Communications Security*, 603–618, 2017.

[42] C. Thapa, P. C. M. Arachchige, S. Camtepe, and L. Sun, "Splitfed: When federated learning meets split learning." In *Proceedings of the AAAI Conference on Artificial Intelligence*, 36(8), 8485–8493, 2022.

[43] C. He, M. Annavaram, and S. Avestimehr, "Group knowledge transfer: Federated learning of large CNNs at the edge." In *The 34th Conference on Neural Information Processing Systems*, Vancouver, BC, Canada, 14068–14080, 2020.

[44] K. He, X. Zhang, S. Ren, and J. Sun, "Deep residual learning for image recognition." In *Proceedings of the IEEE Conference on Computer Vision and Pattern Recognition*, Las Vegas, NV, USA, 770–778, 2016.

[45] D.-J. Han, H. I. Bhatti, J. Lee, and J. Moon, "Accelerating federated learning with split learning on locally generated losses." In *International Workshop on Federated Learning for User Privacy and Data Confidentiality*, Vienna, Austria, 1–12, 2021.

[46] P. Joshi, C. Thapa, S. Camtepe, *et al.*, "Splitfed learning without client-side synchronization: Analyzing client-side split network portion size to overall performance." In *Collaborative European Research Conference*, Cork, Ireland, 1–7, 2021.

[47] K. Chang, N. Balachandar, C. Lam, *et al.*, "Distributed deep learning networks among institutions for medical imaging," *Journal of the American Medical Informatics Association*, 25(8), 945–954, 2018.

[48] G. Hinton, O. Vinyals, and J. Dean, "Distilling the knowledge in a neural network." In *NIPS Deep Learning Workshop*, Montreal, Canada, 1–9, 2014.

Chapter 2

Splitfed learning processing for IoT and Big Data applications

B.C. Soundarya[1], Gururaj H.L.[1], Geetabai S. Hukkeri[2] and S. Priya[1]

Federated learning (FL) and split learning (SL) are two approaches to learning that can greatly enhance ubiquitous intelligence within the Internet of Things (IoT). FL involves combining trained machine learning (ML) models, each using data into a global model. On the one hand SL allows different segments of an ML model to be collaboratively trained on workers in a learning framework. While both FL and SL have their strengths and limitations, they can work together to achieve intelligence in the IoT. Recently, there has been a surge of interest in exploring the synergy between FL and SL making it a key focus area for research. This chapter offers insights into the advancements in FL and SL and provides an overview of cutting-edge technologies for integrating these approaches into an edge computing-based IoT setting. Furthermore, it addresses some existing challenges. The chapter also discusses future research directions to spark further exploration, within the academic community.

Keywords: Big Data; Distributed learning; Federated learning; Split learning; Machine learning; Splitfed learning

2.1 Introduction

The demand for efficient and scalable machine learning (ML) solutions has surged in the rapidly evolving landscape of Internet of Things (IoT) and Big Data [1]. However, traditional ML paradigms face significant challenges when applied to these environments. Constraints such as limited computational resources, bandwidth constraints and privacy concerns hinder their effectiveness. To overcome these challenges, splitfed learning processing has emerged as a promising approach that caters to the unique requirements of IoT and Big Data applications.

[1]Department of Information Technology, Manipal Institute of Technology Bengaluru, Manipal Academy of Higher Education, India

[2]Department of Computer Science and Engineering, Manipal Institute of Technology Bengaluru, Manipal Academy of Higher Education, India

Splitfed learning also called split federated learning is a decentralised way for carrying out ML that takes advantage of what federated learning (FL) can do but also offers solutions to its issues within IoT as well as Big Data domain [2,3]. What this means is that there is no centralisation of data while still training models on numerous devices, thus ensuring confidentiality concerning data while minimising communication costs. However, when it comes to dealing with low resource devices coupled with large amounts of datasets. the conventional FL techniques may have issues associated with communication bandwidth together with computational complexity [4].

To begin with, this chapter serves to set the stage for understanding the development of splitfed learning and its applications in IoT and Big Data. The core ideas, vital components and the motivation behind its incorporation in IoT and Big Data settings will be explored. Additionally, a thorough investigation will be carried out on how splitfed learning is capable of conquering the obstacles presented by traditional FL techniques in these environments by implementing creative approaches.

The chapter will further offer an overview of the splitfed learning applications in IoT and Big Data, detailing different use cases. These are domain-specific requirements that must be satisfied by such solutions as they may vary from predictive maintenance services for industrial IoT devices to real-time analytics on large sensor networks. Besides this, case studies and practical implementations of splitfed learning in a number of real-life situations will be discussed, which would provide an understanding of its effectiveness and possible impact.

It is only after reading this chapter that the reader can claim to have a full understanding of the splitfed learning processing and how it enables efficient, scalable and privacy-protecting ML systems in the context of IoT and Big Data. The topic under consideration in this preliminary review will be used as a reference point for subsequent sections that will delve into the technicalities, algorithms and case studies of splitfed learning processing applications in different areas.

2.2 Fundamentals of splitfed learning

2.2.1 Distributed learning

Amid the era of prolific data and complex analytics, centralisation as the traditional way to ML experiences a number of obstacles. The expansion in data size, velocity and variability presents difficulties for centralised learning systems, causing processing delays with large datasets and requiring high computational power [5]. Moreover, due to an increased demand for privacy and security in the world of data sources, distributed solutions should be created. These solutions would permit education from decentralised bases and guarantee the protection of sensitive information.

Distributed learning appears to be a promising approach to address these challenges through decentralising the learning process over multiple computing nodes/devices. In contrast with the classical [6] centralised learning paradigm with one centre of the computation that receives and processes data for learning, distributed learning aims to exploit the computation power of the distributed computing systems to execute the learning tasks on the data of distributed sources while acknowledging the data locality and privacy [7].

This introductory chapter delves into the fundamentals of distributed learning, with emphasis on the algorithms and the principles to solve problems in modern data analytics. The main ideas of distributed learning explained in this chapter are the distributed optimisation algorithms, communication protocols and the model aggregation techniques, along with the underlying reasons behind the increasing application of distributed learning in a wide range of areas including the IoT, edge computing and FL [8].

Moreover, we present use cases and examples that highlight the value of distributed learning in areas such as healthcare, finance, manufacturing and telecommunications. From improving predictive maintenance in industrial IoT systems to enabling personalised recommendation systems in e-commerce, distributed learning provides flexible and privacy-preserving solutions to the evolving needs of modern data analytics [9,10].

After reading this chapter, readers will have a good understanding of what distributed learning is and why it is important to achieve the three goals of building scalable, effective and privacy-preserving ML algorithms for Big Data analytics. This chapter provides an introduction to distributed learning; the following chapters will describe the details, algorithms and real applications of the distributed learning in different scenarios.

As shown in their research, Arzovs *et al.* [11] proposed this distribution diagrams to demonstrate uses of ML on the edge (refer to Figure 2.1).

The DL is a branch of learning that further divides into two different methods, which include FL and split learning (SL).

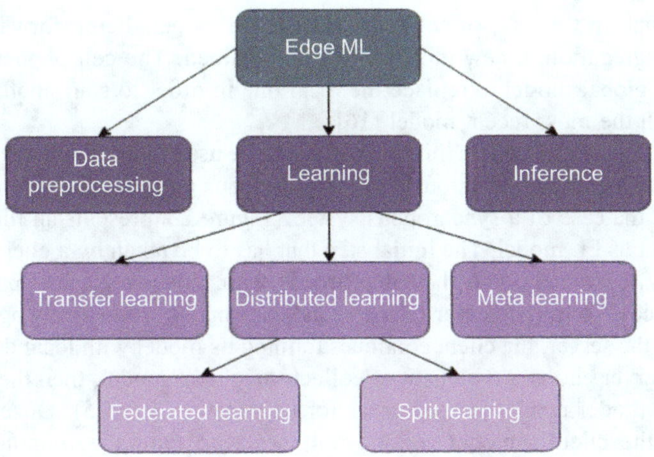

Figure 2.1 Edge machine learning distribution [11]

However, different DL approaches have been developed to counteract those inherent weaknesses of the other methods and to improve the flexibility and distribution of the ML operations over different hardware configurations and over multiple devices. Also, transfer learning and other learning branch methods are commonly used along with the DL at the same time, exploiting their strengths, often for devices with less computational power available than cloud infrastructure.

2.2.2 Federated learning

The initial examined deep learning technique is known as FL. The focus of research in this field primarily revolves around the definition put forth by Google. Occasionally, this approach is referred to as horizontal federated learning (HFL). In this scenario, it is assumed that the clients share their feature spaces. However, this assumption does not always hold true. To address the disparity in data feature space among clients, vertical federated learning (VFL) was developed [12]. A more comprehensive analysis of the various directions of FL can be found in [12], but it is beyond the scope of this study.

The main principle of FL is using ML algorithms on edge devices to strengthen the level of data privacy and security. This is done through keeping the information for each device on its host device. Techniques in use here are differential privacy and secure aggregation. However, some current literature suggests that there is still ambiguity around the requirement to employ these privacy and security approaches in FL [13–15].

Instead of training one ML model at a time, FL allows the training of multiple versions of ML models simultaneously and the synchronisation of their updates into a central model. In this way, communication bandwidth is minimised since only model updates are transmitted rather than the original data used to train each local model [15]. An example use case would be an FL network with mobile devices as participants for a task like keyboard input recognition. Each device trains its model based on keyboard inputs to enhance accuracy in recognising the user's writing style.

After that, in the case of some or all of these models being forwarded to the cloud for aggregation, a new global model is obtained. The cell phones then each receive this global model to replace the local one in order to start another cycle of training with the most recent model [16].

One of the settings where this method could be used is those where the learning procedure can be either synchronous or asynchronous as shown in Figure 2.2. First, let's look at the case of a synchronous system. Figure 2.2 presents an illustration of the synchronous FL model. The initial step that has to be taken by a client (step 1) is to get ready to connect, and then it proceeds with requesting participation in the training process from the server (steps 2 and 3). Finally, after receiving the global model from the server, the client continues training its model with local data (step 4).

If the server chooses the client to collect the global model, then the client will transfer the model results to the server for aggregation (step 5). Once the server aggregates the client's model results, if it needs to continue training, then the training cycle starts again from step 1. If convergence has been achieved in this particular situation, it would be unnecessary to keep on training [17,18].

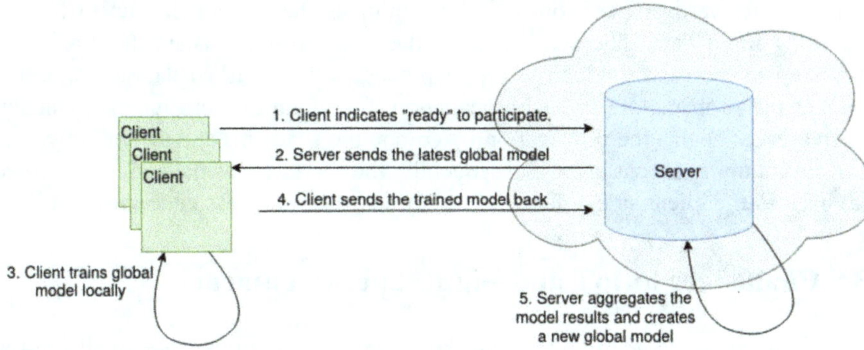

Figure 2.2 Learning procedure [11]

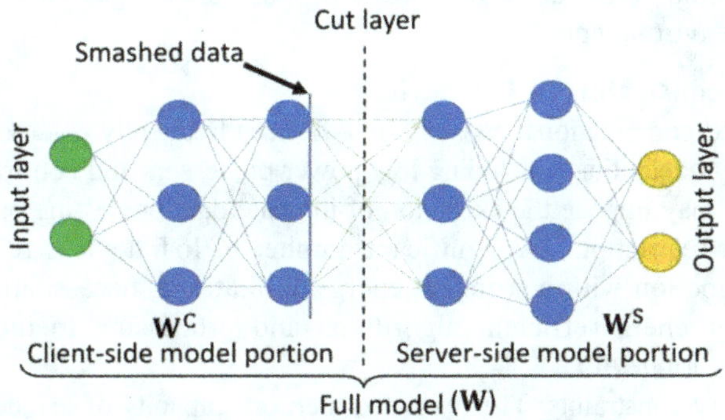

Figure 2.3 Example of split learning

2.2.3 Split learning

In SL, models are partitioned into two or more segments and distributed across the device network. These devices collectively work together to train a single model in a sequential manner, based on how the model is fragmented.

This technique is considered an evolution of FL; however, it has not yet gained widespread popularity. As a result, there are fewer research articles available on existing SL approaches, as well as on privacy and security measures for SL, compared to FL.

Figure 2.3 illustrates an example of SL, where a model is split into two parts. On the left side, there is a client-side segment (*WC*), which could be implemented on devices like Raspberry Pi or other edge or IoT devices. On the right side, there is a server-side segment (*WS*), which could be deployed in the cloud. The division point of the model is referred to as the Cut Layer, and the data shared between the model segments is known as smashed data. The training and inference processes occur sequentially.

During forward propagation, the client initiates the propagation up to the cut layer using input data. The activations of the final layer (smashed data) are then transmitted to the server side. The server processes the smashed data as input and continues the propagation. In backpropagation, following the loss function calculation, the server begins the propagation by computing gradients and activation values for all layers up to the cut layer. Subsequently, the server sends the results (smashed data) back to the client side, where the backpropagation process continues.

2.3 Challenges in IoT and Big Data environments

In the realm of implementing ML and data analytics solutions, specifically in the domains of IoT and Big Data, numerous challenges emerge as a result of the distinctive characteristics inherent in these fields. It is imperative to comprehend and tackle these challenges in order to guarantee the efficacy and efficiency of these systems. Presented below are a few of the primary challenges encountered in IoT and Big Data environments:

1. **Resource constraints in IoT devices**
   - Limited computational capacity: IoT devices frequently possess restricted processing capabilities, including low-power processors and constrained memory, which may impede the execution of intricate data processing operations.
   - Power restrictions: A significant number of IoT devices rely on batteries and function within stringent energy limitations, necessitating the utilisation of energy-efficient algorithms and processing methods to extend battery longevity.
   - Memory constraints: The restricted memory capacity of IoT devices presents obstacles in terms of data storage and processing large quantities of data, underscoring the need for effective data compression and storage strategies.
2. **Bandwidth limitations**
   - IoT devices often function in settings where network bandwidth is restricted, like wireless sensor networks or remote areas with sporadic connectivity. This constraint can hinder the volume of data that can be sent, impacting the effectiveness of data transfer and communication between devices and servers.
   - In IoT environments, applications that require real-time or near-real-time responses necessitate low-latency communication, a goal that can be difficult to meet because of bandwidth constraints and network congestion.
3. **Privacy concerns and data security**
   - Data privacy: The collection of sensitive data, such as personal and location information, by IoT devices has sparked concerns regarding data privacy and the possibility of unauthorised access or misuse.
   - Data encryption: To guarantee the secure transmission and storage of data in IoT environments, it is crucial to employ strong encryption techniques that can safeguard against eavesdropping and unauthorised access.

- Identity management: The management of device and user identities within IoT systems is vital in order to enforce access control policies and prevent unauthorised access to sensitive data.

4. **Scalability issues with Big Data**
   - Data volume: Big Data ecosystems manage enormous amounts of data produced by IoT devices, sensors and various origins, creating obstacles for storage, computation and examination.
   - Scalable computing: Conventional centralised processing systems might not be able to expand effectively to manage extensive data processing operations, leading to the need for distributed computing structures and adaptable algorithms.
   - Data diversity and sophistication: Big Data is distinguished by a range of data formats and arrangements, such as structured, semi-structured and unstructured data, demanding versatile processing methods and algorithms that can manage this diversity of data.

To tackle these obstacles, it is essential to adopt inventive strategies and customised solutions that cater to the unique demands of IoT and Big Data settings. Key techniques like edge computing, FL and distributed processing frameworks are pivotal in surmounting these challenges and facilitating effective and scalable ML and data analytics in IoT and Big Data applications.

Splitfed learning, also known as split federated learning, is a concept in the field of FL that aims to tackle issues related to massive datasets, communication effectiveness and safeguarding privacy. This approach involves dividing the training procedure among various data subsets or devices, allowing for streamlined optimisation and privacy protection, all while harnessing the combined knowledge of decentralised devices or data origins. Here is a summary of the essential elements and methodologies utilised in splitfed learning algorithms:

1. **Communication-efficient optimisation methods:**
   - Quantisation and compression: Techniques to reduce the amount of data exchanged during the training process by quantising model parameters or compressing gradients.
   - Asynchronous methods: Strategies that allow devices to update model parameters asynchronously, reducing the need for synchronisation and communication overhead.
   - Model aggregation techniques: Algorithms for aggregating model updates from different devices while minimising communication costs such as federated averaging.

2. **Partitioning strategies for large-scale data:**
   - Horizontal partitioning involves partitioning data based on samples or rows, where each device or data source contains a portion of the samples. Vertical partitioning, on the other hand, involves partitioning data based on features or columns, with each device containing a subset of features. Hybrid

partitioning, as the name suggests, combines both horizontal and vertical partitioning techniques to achieve an optimal distribution of data across devices.

3. **Privacy-preserving techniques:**
   - Differential privacy: Incorporating random perturbations into the model updates to safeguard against the disclosure of confidential details regarding individual data points.
   - Secure multi-party computation (SMPC): Approaches that enable multiple entities to collaboratively execute a function using their respective inputs while maintaining the confidentiality of those inputs.
   - Homomorphic encryption: A cryptographic method that permits computations to be carried out on encrypted data without the need for decryption, thereby guaranteeing privacy throughout the computation process.
4. **Algorithmic approaches:**
   - Split learning involves dividing the model into two components, one located on the central server and the other on client devices. The only information exchanged between the two components are the model updates, ensuring the privacy of the data.
   - Transfer learning with splitfed utilises transfer learning methods within the splitfed structure to share knowledge from pre-trained models to devices with restricted data, minimising the necessity for extensive local training.

Splitfed learning algorithms aim to facilitate the efficient and privacy-preserving training of ML models on decentralised data sources at a large scale by incorporating these techniques. This makes it an ideal choice for situations where data privacy and communication constraints are of utmost importance.

2.4 Applications of splitfed learning in IoT

1. Healthcare monitoring systems:
   - Scenario: Imagine a network of wearable devices, like smartwatches, that gather health information such as heart rate, blood pressure and activity levels.
   - Application: Splitfed learning allows for the training of models directly on these wearable devices, eliminating the need to send raw health data to a central server. These models can be trained to identify irregularities in health parameters or forecast health-related incidents, all while safeguarding user privacy.
2. Smart home automation:
   - In a smart home setting, a range of IoT devices, including sensors, cameras and smart appliances, produce data concerning occupancy, temperature, energy usage and more. Splitfed learning offers the capability to train models that optimise energy consumption, enhance security by identifying

irregularities in sensor data or personalise user experiences, all without transmitting raw data to a central server. As an example, an edge device can learn occupancy patterns and adapt temperature settings accordingly, leading to improved energy efficiency.

2.5 Applications of splitfed learning in Big Data

1. Personalised recommendations and content filtering:
    - In scenarios where personalised recommendations or content filtering services are offered, splitfed learning can be utilised to train models on user devices without compromising privacy. This approach allows for local model training based on user interaction data, such as viewed videos or liked content, instead of sending sensitive information to a central server. By aggregating model updates, recommendations can be enhanced for all users while still maintaining individual privacy.
2. Healthcare data analysis and disease prediction:
    - In healthcare environments where patient data is confidential and must adhere to privacy laws, splitfed learning facilitates joint model training among separate healthcare facilities while upholding data privacy standards.
    - Hospitals in a specific area gather patient health records, diagnostic images and lab findings. Splitfed learning permits hospitals to locally train disease prediction models using their own patient data. Updates to the models are regularly combined to enhance the precision of disease prediction models throughout the entire network of hospitals without disclosing sensitive patient details.

2.6 Methodology of split federated learning

Split federated learning is an approach that enables the training of ML models using decentralised data, without the need to centralise the data itself. This technique proves to be highly advantageous in situations where data privacy is a priority or when regulatory or technical limitations prevent data centralisation. In split federated learning, the data remains securely stored on its respective devices or servers, while only the model updates are shared between these devices or servers.

1. **Initialisation**
    - At first, a worldwide model is established with arbitrary parameters.
    - Afterwards, the worldwide model is divided into smaller sub-models, each representing a segment of the complete model's structure. These sub-models can be allocated across various devices or servers.
2. **Data splitting**
    - The data for training is divided among numerous devices or servers. Each individual device or server contains a specific portion of the data, which is stored locally.

3. **Local model training**
   - The local sub-model is trained on every device or server using the data available locally. Different ML algorithms like gradient descent or its variations (e.g., stochastic gradient descent, mini-batch gradient descent) can be employed in this training process. Gradients are computed and local model parameters are updated by the device/server during the local training.
4. **Model aggregation**
   - After completing training at the local level, the revised sub-model parameters (gradients) are combined from different devices or servers to create a universal update. Different aggregation techniques, including averaging, weighted averaging or more advanced methods like federated averaging, which take into account the quantity of local data and model effectiveness, can be utilised.
   - Upon finishing training locally, the updated sub-model parameters (gradients) are merged across devices or servers to create a global update. Various aggregation techniques, such as averaging, weighted averaging or more sophisticated methods like federated averaging, which consider the amount of local data and model performance, can be utilised.
5. **Global model update**
   - The updated parameters of the global model are obtained by applying the global update to the global model. These updated parameters are subsequently distributed back to the devices or servers.
6. **Iterative process**
   - The process of steps 3–5 is performed repeatedly for a specific number of rounds or until the convergence criteria are satisfied. During each iteration, devices or servers locally train their models, combine the updates and obtain the latest global model parameters.

Algorithms commonly used in split federated learning include:

1. Various algorithms such as stochastic gradient descent (SGD), mini-batch gradient descent and federated averaging are commonly employed for local model training and global model updating.
2. Federated averaging (FedAvg): The FedAvg algorithm encompasses the process of aggregating gradients or model parameters from numerous devices or servers in order to achieve a global update.
3. Secure aggregation involves the utilisation of techniques like secure multi-party computation (MPC) or homomorphic encryption to ensure the secure aggregation of model updates while maintaining data privacy.
4. Differential privacy is a technique that can be incorporated into the training process to safeguard sensitive information in local data, thereby ensuring privacy.

Split federated learning provides a harmonious blend of data privacy and model accuracy by enabling model training on distributed data while minimising privacy vulnerabilities.

2.7 Performance evaluation and benchmarking in splitfed learning

1. **Metrics for evaluating splitfed learning models**

   Model precision: This parameter evaluates the precision of splitfed learning models on test datasets. It is commonly evaluated using standard metrics like precision, recall, F1-score or area under the receiver operating characteristic (ROC) curve (AUC). Model precision offers valuable insights into the capability of splitfed learning models to make precise predictions on unseen data.

   Communication overhead: Within the context of splitfed learning, communication overhead pertains to the level of communication that is necessary between devices and servers throughout the process of training a model. This encompasses various factors, including the number of communication rounds, the quantity of data transmitted between devices and servers, and the frequency at which model updates occur. The minimisation of communication overhead is crucial for ensuring efficient training in distributed environments that have limited bandwidth and computational resources.

   Privacy preservation: Splitfed learning is designed to maintain the confidentiality of sensitive information while facilitating collaborative model training. The evaluation of privacy preservation involves the use of metrics such as differential privacy guarantees and privacy loss bounds. Differential privacy measures the extent to which splitfed learning models safeguard privacy by assessing the influence of an individual's data on the overall model output. Privacy loss bounds offer theoretical assurances regarding the extent of privacy leakage during model training, thereby safeguarding sensitive data.

   Convergence rate: The rate of convergence refers to the speed at which splitfed learning models reach an optimal solution during the training process. It is evaluated by monitoring the loss function or objective function throughout the training iterations. A higher convergence rate implies that splitfed learning models can attain satisfactory performance with a reduced number of training iterations, thereby minimising computational time and resource usage.
2. **Benchmarking against traditional federated learning**

   Model accuracy: Evaluating the performance of splitfed learning models and traditional FL models requires a comparative analysis of their predictive accuracy on similar datasets. By comparing the accuracy metrics of both splitfed learning and traditional FL methods, researchers can determine the effectiveness of splitfed learning in achieving comparable or superior predictive performance.

   Communication overhead: Evaluating the efficiency gains of splitfed learning in comparison to traditional FL methods through benchmarking the communication overhead enables researchers to analyse factors like communication rounds, data transmission volume and model update frequency.

   Privacy preservation: A thorough examination of the privacy-preserving features of splitfed learning in comparison to conventional FL techniques requires the assessment of metrics like differential privacy assurances or privacy loss boundaries. Through a comparison of the privacy safeguards offered by splitfed learning and traditional FL methods, scholars can gauge the extent of privacy preservation attained by each approach.

Convergence rate: A comparative study on the convergence rate of splitfed learning models in contrast to conventional FL models requires an evaluation of both the speed and stability of convergence. To analyse the convergence patterns of splitfed learning and traditional FL approaches, researchers can monitor the loss function or objective function throughout training iterations and determine how quickly models reach an optimal solution.

3. **Experimental results and comparative analysis**

Real-world datasets: Researchers can assess the robustness and generalisation capabilities of splitfed learning models across various application domains by conducting experiments on real-world datasets. This approach enables them to evaluate the performance of these models in practical scenarios, as real-world datasets possess diverse characteristics.

Simulation studies: Simulation studies allow researchers to evaluate the scalability and resilience of splitfed learning models across different conditions, including various network topologies and data distributions. Through simulating diverse scenarios and analysing the performance of splitfed learning models, researchers can acquire valuable insights into their behaviour and performance attributes in diverse settings.

Comparative analysis: Conducting comparative evaluations between splitfed learning and conventional FL methods across various criteria such as model accuracy, communication overhead, privacy preservation, and convergence rate offers valuable perspectives on the strengths and weaknesses of each technique. Through assessing the efficiency of splitfed learning and traditional FL methods in different dimensions, scholars can pinpoint the respective benefits and compromises of each strategy.

Case studies: By showcasing successful implementations and deployments of splitfed learning models, researchers can demonstrate the potential impact and advantages of splitfed learning in addressing real-world issues. Presenting case studies and use cases that highlight the efficacy of splitfed learning compared to traditional FL methods in different application domains, such as healthcare, finance or IoT, offers valuable insights into the practical implications of splitfed learning.

To summarise, the evaluation of performance and benchmarking in splitfed learning includes the assessment of important metrics like model accuracy, communication overhead, privacy preservation and convergence rate. This involves comparing splitfed learning models with traditional FL methods, conducting experiments on real-world datasets and simulation studies, analysing across various metrics and showcasing case studies to illustrate the effectiveness of splitfed learning in different application areas. Through the use of these thorough evaluation methods, researchers can acquire valuable insights into the performance attributes and real-world implications of splitfed learning in privacy-preserving ML applications.

2.8 Conclusion

FL and SL are two emerging methods of collaborative learning that can work together to enable ubiquitous intelligence in the IoT. This chapter provides an overview of the recent advancements in FL and SL within the context of edge computing-based IoT. Additionally, it presents a survey on the state-of-the-art technologies that combine SL with FL to facilitate ubiquitous intelligence in IoT. The survey highlights the significant progress made in various aspects of the SL-FL combination such as model decomposition methods, techniques for improving performance and efficiency, and privacy protection. However, it is important to note that research in this area is still in its early stages, and there are numerous open issues that require further investigation. The chapter also identifies some of these open problems and discusses potential directions for future research in this emerging field.

References

[1] C. Thapa, P.C.M. Arachchige, S. Camtepe, and L. Sun. "SplitFed: When federated learning meets split learning." In *Proceedings of the AAAI Conference on Artificial Intelligence*, vol. 36, no. 8, pp. 8485–8493, 2022.

[2] P. Joshi, C. Thapa, S. Camtepe, M. Hasanuzzamana, T. Scully, and H. Afli. "SplitFed learning without client-side synchronization: Analyzing client-side split network portion size to overall performance." *arXiv*:2109.09246, 2021, https://doi.org/10.48550/arXiv.2109.09246.

[3] A. Abedi and S.S. Khan. "FedSL: Federated split learning on distributed sequential data in recurrent neural networks." *Multimedia Tools and Applications*, vol. 83, no. 10, pp. 28891–28911, 2024.

[4] M.A. Khan, V. Shejwalkar, A. Houmansadr, and F.M. Anwar. "Security analysis of splitfed learning." In *Proceedings of the 20th ACM Conference on Embedded Networked Sensor Systems*, pp. 987–993, 2022.

[5] H. Chang, V. Shejwalkar, R. Shokri, and A. Houmansadr. "Cronus: Robust and heterogeneous collaborative learning with black-box knowledge transfer." *arXiv*:1912.11279, 2019, https://doi.org/10.48550/arXiv.1912.11279.

[6] D. Pasquini, G. Ateniese, and M. Bernaschi. "Unleashing the tiger: Inference attacks on split learning." In *Proceedings of the 2021 ACM SIGSAC Conference on Computer and Communications Security*, pp. 2113–2129, 2021.

[7] S. Pal, M. Uniyal, J. Park, et al. "Server-side local gradient averaging and learning rate acceleration for scalable split learning." *arXiv*:2112.05929, 2021, https://doi.org/10.48550/arXiv.2112.05929.

[8] H. Madaan, M. Gawali, V. Kulkarni, and A. Pant. "Vulnerability due to training order in split learning." In *ICT Systems and Sustainability: Proceedings of ICT4SD 2021*, Singapore: Springer Nature Singapore, vol. 1, pp. 103–112, 2022.

[9] Z. Zhang, A. Pinto, V. Turina, F. Esposito, and I. Matta. "Privacy and efficiency of communications in federated split learning." *IEEE Transactions on Big Data*, vol. 9, no. 5, pp. 1380–1391, 2023.

[10] J. Wang, H. Qi, A.S. Rawat, et al. "FedLite: Improving communication efficiency in federated split learning," 2022. Available: https://openreview.net/forum?id=VO-HUrkHSY.

[11] A. Arzovs, J. Judvaitis, K. Nesenbergs, and L. Selavo. "Distributed learning in the IoT–edge–cloud continuum." *Machine Learning and Knowledge Extraction*, vol. 6, no. 1, pp. 283–315, 2024.

[12] A. Chopra, S.K. Sahu, A. Singh, et al. "Adaptive split learning." In *Federated Learning Systems (FLSys) Workshop@ MLSys 2023*, 2023.

[13] M. Gawali, C.S. Arvind, S. Suryavanshi, et al. "Comparison of privacy-preserving distributed deep learning methods in healthcare." In *Medical Image Understanding and Analysis, MIUA 2021*. Cham: Springer International Publishing, pp. 457–471, 2021.

[14] X. Lyu, S. Liu, J. Liu, and C. Ren. "Scalable aggregated split learning for data-driven edge intelligence on internet-of-things." *IEEE Internet of Things Magazine*, vol. 6, no. 4, pp. 124–129, 2023.

[15] B. Di Martino, M. Graziano, L.C. Cante, and D. Cascone. "Analysis of techniques for mapping convolutional neural networks onto cloud edge architectures using SplitFed learning method." In *International Conference on Advanced Information Networking and Applications*, Cham: Springer International Publishing, pp. 163–172, 2022.

[16] C. Xu, J. Li, Y. Liu, Y. Ling, and M. Wen. "Accelerating split federated learning over wireless communication networks." *IEEE Transactions on Wireless Communications*, vol. 23, no. 6, pp. 5587–5599, 2024.

[17] W. Zhang, T. Zhou, Q. Lu, Y. Yuan, A. Tolba, and W. Said. "FedSL: A communication-efficient federated learning with split layer aggregation." *IEEE Internet of Things Journal*, vol. 11, no. 9, pp. 15587–15601, 2024.

[18] X. Chen, J. Li, and C. Chakrabarti. "Communication and computation reduction for split learning using asynchronous training." In *2021 IEEE Workshop on Signal Processing Systems (SiPS)*, IEEE, pp. 76–81, 2021.

Chapter 3

Blockchain-driven splitfed learning for data protection in IoT setting

Shampa Rani Das[1], N.Z. Jhanjhi[1], David Asirvatham[1], Farzeen Ashfaq[1], Danish Javed[1] and Gururaj H.L.[2]

Blockchain technology amalgamation seems to be an effective means to confront security and privacy constraints in a wide variety of domains. The proliferation of the Internet of Things (IoT) and the rising prevalence of intelligent surroundings necessitate the protection of highly confidential data and the integrity of operations. Individualized data is susceptible to exploitation and eavesdropping as traditional resource-sharing techniques frequently fail to provide needed anonymity and security. Techniques like splitfed learning (SFL) with blockchain assistance provide a viable way to overcome these obstacles. SFL explores numerous ways that the approaches may be applied regarding IoT, including strengthening data privacy, improving security measures, and fostering confidence in intricate ecosystems. It makes optimal use of the immutable and decentralized properties of blockchain technology to establish an open and secure setting for data exchange. This stimulates collaborative abilities and information exchange in an uncertain accomplishing while also protecting highly confidential information. This research emphasizes the comprehension of how technological advancements are influencing safeguarding information amid today's globalized world by demonstrating the ever-evolving connection between blockchain and SFL. It illustrates how incorporating blockchain technologies is essential to lessening potential hazards and strengthening IoT resiliency which will eventually allow for more trustworthy and secure interactions in this era of digitalization. Blockchain-driven SFL is an indispensable field for subsequent development in subjects of cybersecurity and data privacy as the IoT-enabled infrastructure evolves. It has the potential to contribute an

[1]School of Computer Science (SCS), Taylor's University, Malaysia
[2]Department of Information Technology, Manipal Institute of Technology Bengaluru, Manipal Academy of Higher Education, India

integral part in protecting data between users and smart devices. The projected blockchain-enabled SFL framework is an important stride forward in terms of improving data mining's precision and efficiency while preserving the safety of data in smart environments that involve intricate transactions comprising users and intelligent objects.

Keywords: Blockchain; Splitfed learning; Federated learning; IoT; Data protection; Privacy; Security

3.1 Introduction

Blockchain is a more technologically sophisticated, robust platform that accomplishes decentralized consensus beforehand splitfed learning (SFL) emerges. The prospective benefits of integrating blockchain with SFL for enhanced decentralization and data security are prominent comprehending. Significant advancements have been accomplished concerning decentralized federated learning (FL), certain of them leveraging the blockchain in addition to distinct protocol decentralization. The primary intent is to eliminate the need for a centrally located server for client data collection and training models. Blockchain-driven SFL distributes formulating over a shared network to facilitate privacy while decreasing the minimum processing capacity required for Internet of Things (IoT) equipment to be viable.

3.1.1 Federated learning

FL is a machine learning (ML) approach where a few dispersed devices or nodes coordinate collaboratively to train one model while conserving immediate accessibility to their respective data. This approach of operation is suited for data privacy preservation since solely updated models are distributed and each gadget stores its data, allowing its use conducive in scenarios like smart gadgets, where data confidentiality is of the utmost crucial.

One effective way to accelerate shared data without compromising individual confidentiality is to build FL infrastructure networks within organizations [1]. SFL has emerged as an effective strategy to address data fragmentation and can substantially minimize privacy concerns when employed together with algorithms that protect secrecy. Furthermore, integration with blockchain technology ought to enhance security [2,3] and computing efficiency.

3.1.2 Splitfed learning

SFL [4] exhibits improved computing time and interaction optimization while bolstering data privacy and model robustness. It does this by bringing together the resource productivity of FL with the privacy benefits of split learning (SL). A certain amount of the computing load is transferred from smaller stakeholders to a single central server via SL, while confidentiality issues are dealt with via FL [5].

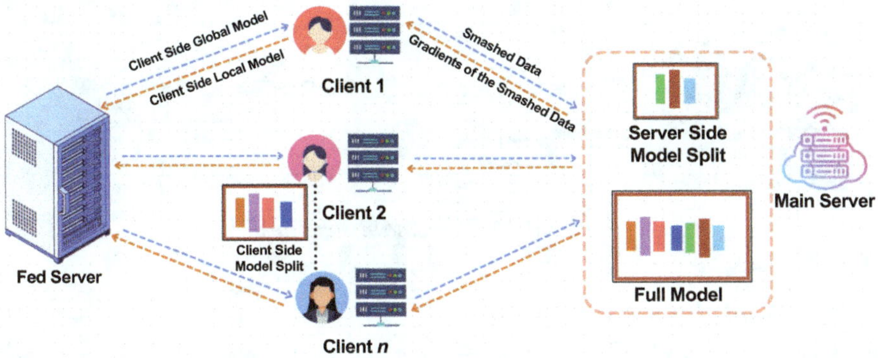

Figure 3.1 Splitfed learning architecture

The SFL architecture is illustrated in Figure 3.1. The following is accomplished by splitting the deep learning (DL) model up into smaller components, enabling decentralized training on local data input without transmitting raw data, and combining only model updates to generate a global model. It protects data privacy while utilizing aggregated information to train high-performing models, making it particularly useful in delicate industries such as healthcare and banking.

3.1.3 Internet of Things

IoT is a network of materially linked items, including structures, automobiles, and additional vehicles, that are implanted with devices with sensors, software, and the connectivity of networks. Several sensors, such as cameras, detecting motion, sensing temperatures, and even more, have been embedded with IoT appliances. These devices gather data, which is then transmitted through wired or wireless connections to central servers or other devices. IoT is widely applicable in numerous distinct businesses. It is employed in smart homes (smart thermostats, security systems), healthcare (wearable fitness trackers, remote patient monitoring), transportation (connected vehicles), agriculture (precision farming), industrial automation (Industry 4.0), and many more. IoT generates enormous volumes of data. This data can be evaluated to get revelations, make real-time decisions, and automate processes. ML and artificial intelligence (AI) are often used to extract valuable information from IoT data.

Although big data and IoT applications have several advantages, the plethora of data they manage makes them extremely vulnerable to cybersecurity attacks, which might impede their development. The authors [6] emphasized the important aspect of comprehending and resolving cybersecurity consequences on emerging technologies by adopting the instance of a medical facility to demonstrate possible dangers and their effect on the generation of big data and IoT applications.

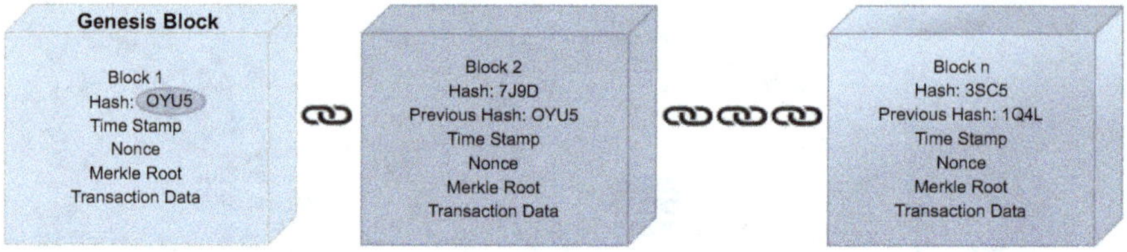

Figure 3.2 Blockchain structure

3.1.4 Blockchain

Blockchain is a distributed and decentralized electronic ledger that renders it possible to record transactions safely and transparently among several entities. It uses cryptographic techniques, fostering trust and minimizing dependency on middlemen to guarantee the immutability and integrity of the recorded data. Transparency (everyone has access to the same version of the ledger) and security (data is cryptographically encrypted and impervious to alteration) are the two main advantages of the technology. In recognition of the likelihood to expedite procedures and improve data quality and provenance, this platform has drawn interest from a variety of sectors.

Blockchain is a digital distributed ledger [7,8] where computers (nodes) share and maintain a chronological record of time-stamped transactions using unique algorithms to achieve consensus, offering advantages like reliability and enhanced computing power due to its decentralized, peer-to-peer architecture [9]. Figure 3.2 illustrates how subsequent blocks of every transaction on the distributed ledger are connected.

Continual inspections are anticipated to develop with the integration of SFL and blockchain into a unified platform for more sophisticated continuous audits. Task automation, which reduces the potential for management influence and unscrupulous conduct, is facilitated by smart contracts.

3.2 Related work

A notable potential for rectifying security and privacy issues in the broader setting of data mining beneath smart contexts with sophisticated contracts is put forward by the incorporation of blockchain technology with SFL. Blockchain-based SFL emphasizes its potential to improve data privacy, fortify security controls, and promote confidence regarding IoT ecosystems, adhering to the evolving dynamics of data protection and safety in the digital age. The literature taxonomy is presented in Figure 3.3.

3.2.1 FL vs SFL

SFL is a variation of FL that incorporates the concept of SL. FL and SFL are both privacy-preserving ML approaches, but they have differences in their core concepts and implementations which are highlighted in Table 3.1. An inventive split-mix FL approach was proposed [10] for the diversity in participants' resources by letting

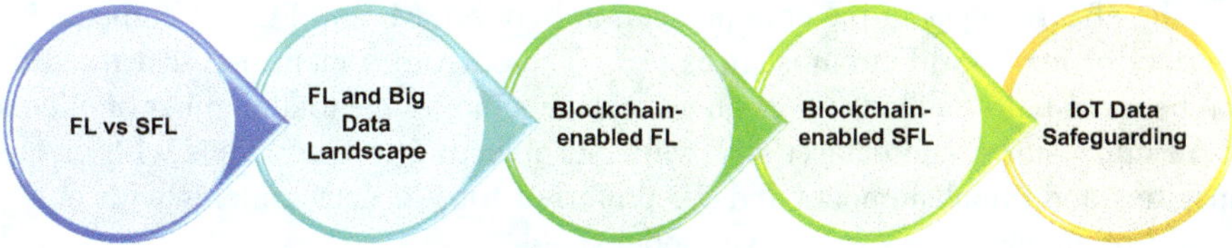

Figure 3.3 Literature taxonomy

Table 3.1 Differences between FL and SFL

	FL	SFL
Data distribution	The data is distributed across multiple devices or nodes, and each device trains a local model using its data	The model is split into parts, and each part is trained locally on different devices or nodes
Model aggregation	It involves aggregating model updates (e.g., weight gradients) from local models to generate a global model. This global model is then shared among the devices for further training rounds	It involves aggregating model parts from local training to reconstruct a global model. The global model is a combination of these model parts
Data privacy	It aims to preserve data privacy by not sharing raw data. Instead, only model updates are shared during the aggregation process	SFL, similar to FL, maintains data privacy by sharing model parts instead of raw data
Use cases	It is well suited for scenarios where data is distributed across devices such as mobile phones, IoT devices, or edge devices. It is often used in applications where privacy is a significant concern such as healthcare or finance	It is more focused on splitting the model for local training while preserving data decentralization. It can also be used in privacy-sensitive applications but with a different approach to preserving privacy
Complexity	It is relatively straightforward in terms of model splitting, with the primary focus on aggregating updates	It introduces the concept of splitting the model, which may add complexity to the model design and aggregation process

in-situ adjustment of model dimensions and resilience through a set of base subnetworks, demonstrating greater effectiveness in communication, storage, and inference in comparison to currently employed heterogeneous-architecture FL methods.

Numerous data distributions and device implementations are used to compare FL and split neural networks (SplitNN), and findings indicate that FL is more effective in terms of communication overhead for IoT scenarios [11]. SplitNN outperforms FL with imbalanced data but lags with extremely not autonomous and precisely the exact dispersed information.

The effectiveness of the communication between SL and FL was compared in a variety of real-world circumstances [12]. The outcomes identified that the decision between both methods relies on variables including the total number of clients, model dimensions, and dataset size, with SL objecting to the models with greater capacities and client numbers and FL preferred for big data samples with only a handful of clients or less extensive model sizes.

SFL can accommodate clients with varying resources and capabilities. Clients can perform computations on their local segments, and the split model design allows flexibility in terms of the complexity and size of individual segments. It is therefore ideally suited for heterogeneous situations, where clients have diverse infrastructure and computational capabilities. It has the potential to lower the training latency by enabling users to do calculations locally and simultaneously. The latter is of major implications in scenarios requiring real-time or nearly immediate model evolves like edge computing and connected objects.

Clients using conventional FL must upload their updated models to a centralized server subsequently every training round. SFL, on the contrary, divides the model into components or different sub-models, and only those upgrades are delivered. It is therefore more appropriate for low-bandwidth or high-latency computer networks as it greatly lessens the transmission load on clients and the centralized server.

SFL is an intriguing approach for addressing latency, confidentiality, and difficulties with communication in distributed ML. It unites the advantageous features of FL and SL. It could potentially enhance the effectiveness and security of collaborative model training, particularly in situations when there are a variety of clients and limited resources in the network. FL and SFL share the identical objective of addressing privacy issues and enable cooperative training for models in decentralized settings.

3.2.2 FL and big data landscape

FL represents one of the ML domains that has experienced significant expansion due to its decentralized data strategy and safeguards for privacy and safety aspects that guarantee compliance with impending client information shielding rules and regulations. FL extends the benefits of ML to less extensive realms when there is insufficient training data to develop an autonomous ML model protecting privacy.

FL is an intriguing approach for distributed computation without disclosing the data itself, but its prevalent model updates and communication requirements can be very taxing, especially for clients with limited resources. As a result, its efficacy for enormous-scale learning of models with local client data may be constrained. The difficulties FL confronts in contexts with limited computer power and poor transmission rates possibly impede its application in real-world settings. FL which collaboratively stimulates a global model among several portable electronic devices has come to prominence considered a potential approach for enormous statistical analysis. Meanwhile, uploaded gradient vectors have the potential to expose participants' sensitive data to an unreliable server.

A privacy-enabled FL technique has been built [13] for the Industrial Internet of Things (IIoT) that is capable of maintaining the anonymity of shareable attributes and local gradients by attaining differential privacy. A consortium blockchain network was incorporated [14] that sought to raise the security level of FL on mobile networks. The centralized server identifies individuals who have substantial esteem values following the reputation measure. A generative adversarial nets (GANs)-based poisoning attack was conducted [15] against FL systems.

Model poisoning differs from poisoning of data in that it directly influences the global model rather than utilizing fictitious data to affect it. Research has shown that model-poisoning occurrences have a greater probability of being effective than data-poisoning threats [16,17]. Having an extensive FL substance at a location with lots of customers appears to increase the efficacy of model poisoning threats. The global model can generally be conveniently poisoned in poisoning a model attempt because the attacker can manipulate the latest version before transmitting it to the centralized server for aggregation.

Attacks that manipulate training datasets consist of including patterns that mislead ML models or combining classes to trick them [18,19]. The attacks in FL case scenarios featuring data insertion and alteration might be classified as ML data poisoning attacks. A few recent FL preventative measures and defensive strategies are presented in Table 3.2.

A privacy-preserving FL scheme that enables to perform operations on encoded information without exposing the contents of the data, grounded on cumulatively homomorphic cryptography [24]. A distributed specific stochastic gradient descent approach was used to accomplish distributed encryptions and lower expenditures associated with communications.

Pruning lessens the model size, enabling more efficient handling of backdoor attacks and communication bottlenecks while not requiring the sharing of a full-fledged model. Anomaly detection techniques are employed in FL to identify deviations from normal behavior and detect various attacks like data poisoning, model poisoning, and Trojan threats. These methods, including clustering, Euclidean distance, autoencoders, and loss-based rejection, are applied to distinguish malicious client updates and enhance the security of FL systems.

The Sniper approach was proposed to distinguish authorized clients and significantly reduce the efficacy of poisoning operations, even in scenarios with some invaders, addressing the uncertainty of the effectiveness of distributed poisoning compared to single-attacker scenarios. Knowledge distillation is a model

Table 3.2 FL defense techniques against threats mitigations

Authors	**Year**	**Threats**	**Defensive strategies**
Jiang *et al.* [20]	2022	Backdoor attacks model	Pruning
Ito *et al.* [21]	2020	Update poisoning	Anomaly detection
Cao *et al.* [22]	2019	Poisoning	Sniper
Li and Wang [23]	2019	Eavesdropping inference GANs	Knowledge distillation

compression technique that transfers information from a neural network that has been entirely trained to a smaller model, reducing computational costs. A federated model distillation framework was utilized to leverage the concept to enhance client data security by sharing knowledge rather than model parameters, allowing for individualized ML models and the usage of processors to compile and distribute knowledge among clients.

3.2.3 Blockchain-enabled FL

Blockchain is used in an FL setting to assure immutability, encourage customer interaction, and allow safe uploading and monitoring of model modifications [25]. A consensus committee verifies and scores local model modifications before uploading the fresh global model to the blockchain network, which substitutes the main server in centralized FL with smart contracts.

Implementing blockchain-enabled FL [26,27] improves the data safety of IoT ecosystems by addressing issues related to data privacy and security that arise from centralized storage and computing in IoT applications. An FL framework for the IIoT is built on the blockchain and reinforced by Intel Software Guard Extension (SGX) technology [28]. For improving the privacy and resiliency of structure sharing and function in IIoT, the architecture securely aggregates local models, inspects their provenance, makes certain model trustworthiness through consensus techniques, and safeguards in a secure manner the global model on the distributed ledger by employing SGX-enabled processors within blockchain nodes.

An approach was developed [29] for overcoming information confidentiality difficulties and scalability problems in mobile-edge computational associations, facilitating secured and decentralized AI trained models on mobile gadgets. Decentralized, safeguarding confidentiality FL is gaining popularity [30] as a way to supplement conventional model-driven design in wireless transmissions, especially in light of the advent of 5G networks.

The expanding potential of FL in mobile edge networks was assessed [31] as an effective strategy for privacy concerns, noting difficulties with transmission expenses, allocating resources, and security. Authors presented the notion of FL-enabled intelligent fog radio networks of accessibility [32], which would use fog computing and AI at the network edge to overcome constraints with data offloading and trained model expenditures.

The proliferation of FL is discussed [33] as an ML paradigm that alleviates privacy concerns by enabling local model training on the end-user's infrastructure while considering the privacy and security hurdles and providing simulation results illustrative of the aforementioned issues and potential remedies.

FL was outlined as a distributed ML method that protects privacy [34], stressing the technique's significance and feasible integration into emerging automotive IoT infrastructures such as cooperative autonomous vehicles and intelligent modes of transportation. A comprehensive analysis of FL's security and privacy aspects was provided [35] highlighting security threats like communication bottlenecks and poisoning while emphasizing the vitality of protecting privacy against inference-based attacks.

A trustworthy technique for IIoT anomaly detection was developed [36] incorporating FL and deep reinforcement learning, improving privacy protection, and attaining high precision in identifying abnormalities across various applications. An anomaly detection paradigm was proposed [37] for industrial controls in automated manufacturing that makes use of FL to increase detection accuracy and minimize bandwidth usage while retaining attainable edge computing hardware restrictions.

A scenario that combines blockchain with FL was presented [38] to tackle network congestion problems, proposing client selection based on both local learning accuracy and participation frequency, and demonstrating through simulations that this approach leads to quicker convergence and increased stability. The privacy hazards connected to centralized sensor data for ML and FL were explored [39] to preserve user data privacy while mitigating shortcomings with communication expenses, diverse data circumstances, and extra privacy controls in the wider setting of IoT applications.

3.2.4 Blockchain-enabled SFL

Digitized twin-capable asynchronous SFL is proposed [40] to enhance confidentiality and communication efficacy in e-healthcare platforms with the IIoT, demonstrating its effectiveness through experiments. RingSFL [41] is a distributed learning scheme that optimizes FL by factoring into individual user heterogeneity, boosting data privacy using model partitioning that blends in a ring topology.

SFL comprises extensive analysis and experiments elaborated [42] that untargeted and distance-based strategies for data poisoning incidents have a more pronounced effect in evading classifier outcomes when contrasted with targeted attacks. SplitFed's lower-dimensional model updates made it more resilient to substantial model poisoning attempts than conventional FL [43,44]. It mitigates privacy issues linked to centralized ML approaches by enabling clients to train and evaluate ML models without revealing the underlying data [45].

Blockchain-based SFL [46] facilitates collaboratively inclined, privacy-conscious training of models to feed the IoT data while offloading computing responsibilities from clients with limited resources, achieving 86% accuracy and evolved decentralization.

SFL is a distributed training method where users submit intermediate concentrations to a centralized server rather than raw data. A framework was proposed [47] that is model inversion (MI) resistant during training through employing an attacker-aware feature extractor and attaining high reconstruction error rates and bare necessity accuracy decrease on the dataset to counter MI attacks on SFL.

3.2.5 IoT data safeguarding

IoT, the interconnected network of smart devices, offers benefits like automation and efficient information flow but is susceptible to cyber threats, including ransomware attacks, which pose significant risks to its functionality and security [48–51]. The incorporation of blockchain technology can strengthen security in IoT applications whilst overcoming several issues that exist, including latency, energy efficiency, and privacy [52]. Amalgamating blockchain with IoT offers significant ways to fortify system security and privacy powered by the IoT.

3.3 Proposed framework

The proposed framework utilizes blockchain-driven SFL to enhance data mining precision, reinforce information privacy and security in intelligent spaces, and demonstrate the potential of SFL methodologies for strengthening data safeguarding. Figure 3.4 depicts the proposed blockchain-based SFL structure for data safety.

The proposed framework comprises three key layers, the IoT layer, the SFL layer, and the blockchain layer. The blockchain layer encompasses various components, including smart contracts [53], consensus mechanisms, data safeguarding measures, encryption, access control, auditing, and anomaly detection, all of which contribute to ensuring the confidentiality, integrity, and controlled access of sensitive data in the SFL process.

Meanwhile, the SFL layer focuses on the core of SFL, enabling multiple stakeholders to cooperatively train an ML model while preserving data decentralization and privacy, with client model splitting and secure model aggregation being central to this process. It ensures that sensitive data remains on the client's side and is never shared directly, making SFL a valuable technique for scenarios where data confidentiality is crucial.

The stated framework's incorporation of the IoT idea adds the dimension of intricacy and ingenuity to the ecosystem. A great deal of data is produced in instantaneous mode by IoT devices which comprise a variety of networked sensors, actuators, and smart gadgets. Diverse sectors may undergo radical change as a result of the enlargement of IoT, blockchain, and SFL layers.

3.3.1 IoT layer

The deployment of *IoT devices*, such as environmental sensors, allows for the collection of data on environmental factors including temperature, humidity, air quality, and pollution levels. A stream of unprocessed sensor readings is produced by these sensors as they constantly accumulate data in real time.

Sensor data encompasses unprocessed readings that are frequently presented as measurements, sensor values, or data points. These data can be in many forms and might contain timestamps, sensor IDs, and numbers. To a central data intake point, sensor data can be sent wirelessly or through connected connections.

The *data processing* layer receives the sensor data and handles and organizes it in preparation for further analysis. Data processing involves various crucial activities, including getting rid of or fixing any abnormalities, outliers, or inaccuracies in the sensor measurements is known as data cleaning. To guarantee that all sensor values are on the same scale and to facilitate proper analysis, data must be normalized. Limiting the amount of data and improving analytical efficacy can be achieved by summarizing or aggregating data across predetermined periods

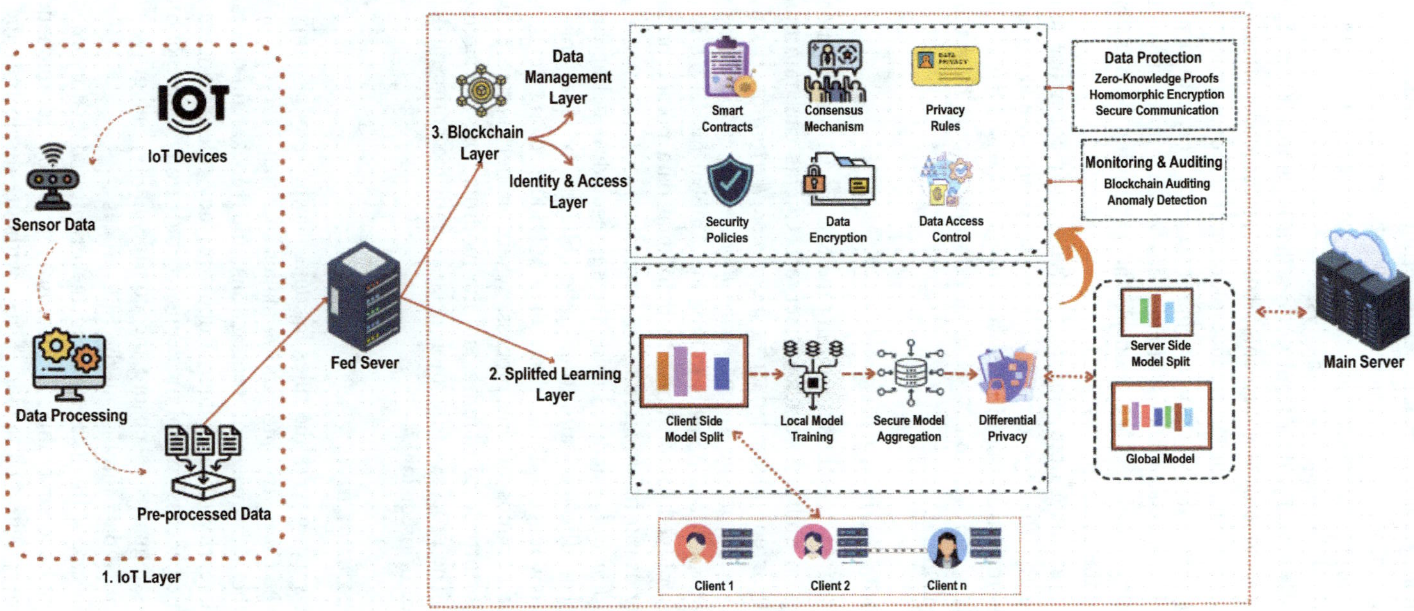

Figure 3.4 Proposed blockchain-driven SFL in IoT setting for data protection framework

(e.g., on an hourly basis). Feature engineering is the procedure of formulating attributes that can be more instructive for subsequent analysis. The data is transformed into a more structured and feasible form after these processing steps, making it ready to be subjected to analysis.

Preprocessed data is data from sensors that have undergone feature enhancement, normalization, cleaning, and compilation. Its structure offers possibilities for analysis, modeling, and the emergence of facts. Preprocessing contributes to the overall data quality, allowing it to be appropriate for ML, statistical evaluation, or illustration. Preprocessed data is often used for functions such as establishing ecological metrics, training ML models, and generating reports for users in the latter stages of the system.

Capturing raw data from sensors, verifying their coherence and correctness, and compiling the findings for evaluation and decision-making are all that are required in the transition from IoT devices to preprocessed data. Data transformation is a fundamental step to extract required valuable data from the data that IoT sensors have obtained.

3.3.2 Splitfed learning layer

Client model splitting comprises splitting the ML model into segments that can be trained locally on multiple devices or servers. Clients can train their model segments using their confidential data in *local model training*. Highly confidential information is kept private since *secure model aggregation* safely merges the locally trained model components to produce a global model. *Differential privacy* [54] can be employed to add noise to individual data points, protecting privacy while allowing useful information to be extracted.

The *global model* is updated using the aggregated model updates which is the target model that all clients aim to improve collectively. The global model incorporates the knowledge gained from the individual client updates. SFL often involves several rounds of this process, where model parts are redistributed to clients, they perform local training, and the modifications are aggregated into the global model. The process of iteration persists until the entire model achieves the targeted degree of precision.

3.3.3 Blockchain layer

Smart contracts can govern the rules for data sharing, model training, and reward distribution. The *Consensus mechanism* makes sure that everyone on the collective network on the legitimacy of transactions and blockchain upgrades. It helps maintain the integrity of the ledger. Robust privacy rules, stringent security policies, fine-grained access control mechanisms, and strong data encryption protocols are essential safeguards to ensure the confidentiality, integrity, and controlled access of sensitive data throughout the SFL process.

Data provenance involves tracking the origin and history of data, ensuring transparency and accountability. *Zero-knowledge proofs* let one side demonstrate their familiarity with particular facts without disclosing the data itself.

Homomorphic encryption [55] is a cryptographic method enabling functions to be carried out with encrypted data without having decryption. *Secure communication* ensures that data transmission between participants is encrypted and secure.

Blockchain auditing monitors transactions and activities for transparency and compliance. A blockchain system may preserve the local and global model hashes on-chain to facilitate auditability. *Anomaly detection* identifies unusual or unauthorized activities in the system.

Overall, it underscores the synergy among IoT, blockchain technology, and SFL, emphasizing data protection, secure collaboration, and transparency, making it particularly valuable in domains where data privacy and secrecy are the ultimate.

3.3.4 Case study

The IoT has revolutionized the acquisition of data in our progressively networked world by facilitating surveillance of numerous ecological parameters through an interconnected set of sensors. However, the massive volumes of private data that IoT devices produce raise pertinent issues regarding data security and privacy. An unprecedented framework is being put into place to address these issues. This innovative approach's practical application is explored in the case study that follows.

A local government's objective in implementing a smart city initiative is to gather and scrutinize data from different IoT devices placed everywhere in the city. These gadgets, which produce a ton of data, include surveillance cameras and traffic monitors. Preeminent considerations pertain to the anonymity of information, security, and adherence to privacy legislation.

The municipality must secure private details like driving license numbers for the sake of maintaining people's privacy. The integrity of data in smart communities has been seriously jeopardized by cyberattacks and data thefts. Aggregated and anonymized data must be accessible to justice officials and urban planners to boost city amenities, traffic control, and safety protocols.

3.3.5 Potential implementation

1. **IoT layer:**
   A network of IoT devices, such as environmental gauges, traffic surveillance systems, and public transit trackers, can be implemented in a city with intelligence. Instantaneous information on traffic circulation, climate, atmospheric pollution, and automobile movements is continually generated by these sensors. Data preprocessing verifies that the data gathered is accurate and consistent.
2. **SFL layer:**
   It can put SFL into practice to allow for safe and private data analysis.
   - Client model splitting: Segments are composed of the ML models used for environmental surveillance and traffic enforcement.
   - Local model training: Authorized users can train their model segments featuring private data such as streaming from traffic cameras and sensor readings.
   - Secure model aggregation: The localized trained model splits are seamlessly aggregated on the blockchain to generate a global traffic surveillance model

without disclosing data indications. Specific data points are safeguarded using privacy approaches such as differential privacy.

- **Blockchain layer:**
  Blockchain technology could eventually be customized to offer a readily available trustworthy platform for exchanging data. Privacy requirements are adhered to via smart contracts, which regulate data access with legislation. Consensus mechanisms preserve data integrity and legitimize data transfers. Highly secure encryption, solid authentication, zero-knowledge proofs, and blockchain auditing processes preserve sensitive data. Data provenance tracks the genesis and trajectory of data, ensuring openness and accountability.

3.3.6 High-level technical specifications

1. **Hardware setup:**
   - IoT equipment can be installed in a specified smart city region.
   - Servers or edge devices can be installed then so that one may train and aggregate models locally.
2. **Software setup:**
   - Authorized municipalities can construct new ML models or utilize pre-existing ones for traffic surveillance.
   - Blockchain nodes can be deployed to validate transactions and record data.
3. **IoT layer's data collection and preprocessing:**
   - Data is persistently collected by IoT devices (e.g., car counts that do not follow traffic rules).
   - Preprocessing is carried out on unprocessed sensor data.
     - Cleaning: Eliminate outliers and abnormalities.
     - Normalization: Make certain that the data scales are consistent.
     - Aggregation: Data would be compiled into a summary (e.g., hourly averages).
4. **SFL layer:**
   - Segment the ML models into partitioning that are appropriate for local training using the client model splitting technique.
   - Local entities use private datasets to train the local model segments.
   - The combination of local model updates while keeping each data point undetectable can be accomplished using secure aggregation techniques. Differential privacy is upheld to conserve individual data.
5. **Blockchain layer:**
   - Smart contract execution specifies regulations for data exchange, access, and cooperation.
   - The consensus mechanism demonstrates data integrity alongside transaction verification and validation.
   - All the required security measures (e.g., cryptography) can be implemented.

6. **Demonstration:**
   - IoT devices identify a traffic concern.
   - Sensitive information is not shared when local models analyze event data.
   - Authorized municipal authorities can view aggregated findings on an interactive dashboard.
7. **Evaluation:**
   - Velocity of data processing
   - Model precision
   - Resilience of security against potential vulnerabilities

The deployment of the proposed framework into the smart city's infrastructure can demonstrate its feasibility and productivity. The city successfully upholds data privacy, security, and legality while stimulating cooperative data analysis to enhance public safety and services. This instance can set guidelines for data security in global endeavors surrounding urban intelligence.

3.4 Conclusion

SFL facilitates interaction among dispersed devices and portrays improved data privacy and computational performance. The amalgamation of blockchain technology with SFL presents an intriguing path for additional investigation and practical deployment as emerging innovations continue to reshape the privacy and security landscape format. Through the integration of various state-of-the-art paradigms, the research endeavor is designed to encourage the development of strong data protection policies in an evolving world that is becoming more networked and data driven. The inclusion of IoT into the proposed framework contributes to automated features. Blockchain, SFL, and IoT convergence offer optimized data security, privacy, and actionable intelligence.

3.5 Future work

It needs to be done to accelerate the practical implementations of blockchain-driven SFL in IoT settings. Additionally, scalability issues must be addressed, and robust privacy-preserving mechanisms must be investigated in diverse use case scenarios.

References

[1] Yang, Q., Liu, Y., Chen, T., and Tong, Y. (2019). Federated machine learning: Concept and applications. *ACM Transactions on Intelligent Systems and Technology (TIST)*, 10(2), 1–19.

[2] Ashraf, H., Hanif, M., Ihsan, U., Al-Quayed, F., Humayun, M., and Jhanjhi, N. Z. (2023). A secure and reliable supply chain management approach integrated with IoT and Blockchain. In *2023 International Conference on*

Business Analytics for Technology and Security (ICBATS) (pp. 1–9). Piscataway, NJ: IEEE.

[3] Humayun, M., Jhanjhi, N. Z., Hamid, B., and Ahmed, G. (2020). Emerging smart logistics and transportation using IoT and blockchain. *IEEE Internet of Things Magazine*, 3(2), 58–62.

[4] Thapa, C., Arachchige, P. C. M., Camtepe, S., and Sun, L. (2022). Splitfed: When federated learning meets split learning. In *Proceedings of the AAAI Conference on Artificial Intelligence* (vol. 36, no. 8, pp. 8485–8493).

[5] Ma, S., Nie, J., Kang, J., *et al.* (2022). Privacy-preserving anomaly detection in cloud manufacturing via federated transformer. *IEEE Transactions on Industrial Informatics*, 18(12), 8977–8987.

[6] Alferidah, D. K., and Jhanjhi, N. Z. (2020). Cybersecurity impact over big-data and IoT growth. In *2020 International Conference on Computational Intelligence (ICCI)* (pp. 103–108). Piscataway, NJ: IEEE.

[7] Das, S. R., Jhanjhi, N. Z., Asirvatham, D., Ashfaq, F., and Abdulhussain, Z. N. (2023). Proposing a model to enhance the IoMT-based EHR storage system security. In *International Conference on Mathematical Modeling and Computational Science* (pp. 503–512). Singapore: Springer Nature.

[8] Patil, P., Sangeetha, M., and Bhaskar, V. (2021). Blockchain for IoT access control, security and privacy: A review. *Wireless Personal Communications*, 117, 1815–1834.

[9] Kumar, M. S., Vimal, S., Jhanjhi, N. Z., Dhanabalan, S. S., and Alhumyani, H. A. (2021). Blockchain based peer to peer communication in autonomous drone operation. *Energy Reports*, 7, 7925–7939.

[10] Hong, J., Wang, H., Wang, Z., and Zhou, J. (2022). Efficient split-mix federated learning for on-demand and in-situ customization. arXiv:2203.09747, https://doi.org/10.48550/arXiv.2203.09747.

[11] Gao, Y., Kim, M., Abuadbba, S., *et al.* (2020). End-to-end evaluation of federated learning and split learning for internet of things. *International Symposium on Reliable Distributed Systems (SRDS)*, Shanghai, China, pp. 91–100.

[12] Singh, A., Vepakomma, P., Gupta, O., and Raskar, R. (2019). Detailed comparison of communication efficiency of split learning and federated learning. arXiv:1909.09145, https://doi.org/10.48550/arXiv.1909.09145.

[13] Hao, M., Li, H., Luo, X., Xu, G., Yang, H., and Liu, S. (2019). Efficient and privacy-enhanced federated learning for industrial artificial intelligence. *IEEE Transactions on Industrial Informatics*, 16(10), 6532–6542.

[14] Kang, J., Xiong, Z., Niyato, D., Zou, Y., Zhang, Y., and Guizani, M. (2020). Reliable federated learning for mobile networks. *IEEE Wireless Communications*, 27(2), 72–80.

[15] Zhang, J., Chen, J., Wu, D., Chen, B., and Yu, S. (2019). Poisoning attack in federated learning using generative adversarial nets. In *2019 18th IEEE International Conference on Trust, Security and Privacy in Computing and Communications/13th IEEE International Conference on Big Data Science and Engineering (TrustCom/BigDataSE)* (pp. 374–380). Piscataway, NJ: IEEE.

[16] Bagdasaryan, E., Veit, A., Hua, Y., Estrin, D., and Shmatikov, V. (2020). How to backdoor federated learning. In *Proceedings of the Twenty Third International Conference on Artificial Intelligence and Statistics*, Palermo, Italy, PMLR, vol. 108, pp. 2938–2948.

[17] Bhagoji, A. N., Chakraborty, S., Mittal, P., and Calo, S. (2019). Analyzing federated learning through an adversarial lens. In *Proceedings of the 36th International Conference on Machine Learning*, Long Beach, CA, PMLR, vol. 97, pp. 634–643.

[18] Fang, M., Cao, X., Jia, J., and Gong, N. (2020). Local model poisoning attacks to {Byzantine-Robust} federated learning. In *29th USENIX Security Symposium (USENIX Security 20)* (pp. 1605–1622).

[19] Nasr, M., Shokri, R., and Houmansadr, A. (2019). Comprehensive privacy analysis of deep learning: Passive and active white-box inference attacks against centralized and federated learning. In *2019 IEEE Symposium on Security and Privacy (SP)* (pp. 739–753). Piscataway, NJ: IEEE.

[20] Jiang, Y., Wang, S., Valls, V., *et al.* (2023). Model pruning enables efficient federated learning on edge devices. *IEEE Transactions on Neural Networks and Learning Systems*, 34(12), 10374–10386.

[21] Ito, R., Tsukada, M., and Matsutani, H. (2021). An on-device federated learning approach for cooperative model update between edge devices. *IEEE Access*, 9, 92986–92998.

[22] Cao, D., Chang, S., Lin, Z., Liu, G., and Sun, D. (2019). Understanding distributed poisoning attack in federated learning. In *2019 IEEE 25th International Conference on Parallel and Distributed Systems (ICPADS)* (pp. 233–239). Piscataway, NJ: IEEE.

[23] Li, D., and Wang, J. (2019). Fedmd: Heterogenous federated learning via model distillation. arXiv preprint arXiv:1910.03581.

[24] Zhang, J., Chen, B., Yu, S., and Deng, H. (2019). PEFL: A privacy-enhanced federated learning scheme for big data analytics. In *2019 IEEE Global Communications Conference (GLOBECOM)* (pp. 1–6). Piscataway, NJ: IEEE.

[25] Li, Y., Chen, C., Liu, N., Huang, H., Zheng, Z., and Yan, Q. (2020). A blockchain-based decentralized federated learning framework with committee consensus. *IEEE Network*, 35(1), 234–241.

[26] Issa, W., Moustafa, N., Turnbull, B., Sohrabi, N., and Tari, Z. (2023). Blockchain-based federated learning for securing internet of things: A comprehensive survey. *ACM Computing Surveys*, 55(9), 1–43.

[27] Zhu, J., Cao, J., Saxena, D., Jiang, S., and Ferradi, H. (2023). Blockchain-empowered federated learning: Challenges, solutions, and future directions. *ACM Computing Surveys*, 55(11), 1–31.

[28] Kalapaaking, A. P., Khalil, I., Rahman, M. S., Atiquzzaman, M., Yi, X., and Almashor, M. (2022). Blockchain-based federated learning with secure aggregation in trusted execution environment for internet-of-things. *IEEE Transactions on Industrial Informatics*, 19(2), 1703–1714.

[29] Nguyen, D. C., Ding, M., Pham, Q. V., *et al.* (2021). Federated learning meets blockchain in edge computing: Opportunities and challenges. *IEEE Internet of Things Journal*, 8(16), 12806–12825.

[30] Niknam, S., Dhillon, H. S., and Reed, J. H. (2020). Federated learning for wireless communications: Motivation, opportunities, and challenges. *IEEE Communications Magazine*, 58(6), 46–51.

[31] Lim, W. Y. B., Luong, N. C., Hoang, D. T., *et al.* (2020). Federated learning in mobile edge networks: A comprehensive survey. *IEEE Communications Surveys & Tutorials*, 22(3), 2031–2063.

[32] Zhao, Z., Feng, C., Yang, H. H., and Luo, X. (2020). Federated-learning-enabled intelligent fog radio access networks: Fundamental theory, key techniques, and future trends. *IEEE Wireless Communications*, 27(2), 22–28.

[33] Ma, C., Li, J., Ding, M., *et al.* (2020). On safeguarding privacy and security in the framework of federated learning. *IEEE Network*, 34(4), 242–248.

[34] Du, Z., Wu, C., Yoshinaga, T., Yau, K. L. A., Ji, Y., and Li, J. (2020). Federated learning for vehicular internet of things: Recent advances and open issues. *IEEE Open Journal of the Computer Society*, 1, 45–61.

[35] Mothukuri, V., Parizi, R. M., Pouriyeh, S., Huang, Y., Dehghantanha, A., and Srivastava, G. (2021). A survey on security and privacy of federated learning. *Future Generation Computer Systems*, 115, 619–640.

[36] Wang, X., Garg, S., Lin, H., *et al.* (2021). Toward accurate anomaly detection in industrial internet of things using hierarchical federated learning. *IEEE Internet of Things Journal*, 9(10), 7110–7119.

[37] Huong, T. T., Bac, T. P., Long, D. M. *et al.* (2021). Detecting cyberattacks using anomaly detection in industrial control systems: A federated learning approach. *Computers in Industry*, 132, 103509.

[38] Kim, Y. J., and Hong, C. S. (2019). Blockchain-based node-aware dynamic weighting methods for improving federated learning performance. In *2019 20th Asia-Pacific Network Operations and Management Symposium (APNOMS)* (pp. 1–4). Piscataway, NJ: IEEE.

[39] Briggs, C., Fan, Z., and Andras, P. (2021). A review of privacy-preserving federated learning for the Internet-of-Things. In Rehman, M.H. and Gaber, M.M (eds.), *Federated Learning Systems: Towards Next-Generation AI*, pp. 21–50, Cham: Springer.

[40] Stephanie, V., Khalil, I., and Atiquzzaman, M. (2023). Digital twin enabled asynchronous splitfed learning in e-healthcare systems. *IEEE Journal on Selected Areas in Communications*, 41(11), 3650–3661.

[41] Shen, J., Cheng, N., Wang, X., *et al.* (2023). RingSFL: An adaptive split federated learning towards taming client heterogeneity. *IEEE Transactions on Mobile Computing*, 23(5), 5462–5478.

[42] Ismail, A. T. Z., and Shukla, R. M. (2023). Analyzing the vulnerabilities in splitfed learning: Assessing the robustness against data poisoning attacks. arXiv:2307.03197, https://doi.org/10.48550/arXiv.2307.03197.

[43] Khan, M. A., Shejwalkar, V., Houmansadr, A., and Anwar, F. M. (2022). Security analysis of splitfed learning. In *Proceedings of the 20th ACM Conference on Embedded Networked Sensor Systems* (pp. 987–993).

[44] Joshi, P., Thapa, C., Camtepe, S., Hasanuzzamana, M., Scully, T., and Afli, H. (2021). Splitfed learning without client-side synchronization: Analyzing

client-side split network portion size to overall performance. arXiv:2109. 09246, https://doi.org/10.48550/arXiv.2109.09246.

[45] Gajbhiye, S., Singh, P., and Gupta, S. (2022). Data poisoning attack by label flipping on SplitFed learning. In *International Conference on Recent Trends in Image Processing and Pattern Recognition* (pp. 391–405). Cham: Springer Nature.

[46] Batool, Z., Zhang, K., Zhu, Z., Aravamuthan, S., and Aivodji, U. (2022). Block-FeST: A blockchain-based federated anomaly detection framework with computation offloading using transformers. In *2022 IEEE 1st Global Emerging Technology Blockchain Forum: Blockchain & Beyond (iGET-blockchain)* (pp. 1–6). Piscataway, NJ: IEEE.

[47] Li, J., Rakin, A. S., Chen, X., He, Z., Fan, D., and Chakrabarti, C. (2022). Ressfl: A resistance transfer framework for defending model inversion attack in split federated learning. In *Proceedings of the IEEE/CVF Conference on Computer Vision and Pattern Recognition* (pp. 10194–10202).

[48] Menon, S., Anand, D., Kavita, *et al.* (2023). Blockchain and machine learning inspired secure smart home communication network. *Sensors*, 23 (13), 6132.

[49] Humayun, M., Jhanjhi, N. Z., Alsayat, A., and Ponnusamy, V. (2021). Internet of things and ransomware: Evolution, mitigation and prevention. *Egyptian Informatics Journal*, 22(1), 105–117.

[50] Kok, S., Abdullah, A., Jhanjhi, N., and Supramaniam, M. (2019). Ransomware, threat and detection techniques: A review. *International Journal of Computer Science and Network Security*, 19(2), 136.

[51] Almusaylim, Z. A., Alhumam, A., and Jhanjhi, N. Z. (2020). Proposing a secure RPL based internet of things routing protocol: A review. *Ad Hoc Networks*, 101, 102096.

[52] Alamri, M., Jhanjhi, N. Z., and Humayun, M. (2019). Blockchain for Internet of Things (IoT) research issues challenges and future directions: A review. *International Journal of Computer Science and Network Security*, 19(1), 244–258.

[53] Amir Latif, R. M., Hussain, K., Jhanjhi, N. Z., Nayyar, A., and Rizwan, O. (2020). A remix IDE: Smart contract-based framework for the healthcare sector by using blockchain technology. *Multimedia Tools and Applications*, 81, 1–24.

[54] Wei, K., Li, J., Ding, M., *et al.* (2020). Federated learning with differential privacy: Algorithms and performance analysis. *IEEE Transactions on Information Forensics and Security*, 15, 3454–3469.

[55] Zhang, L., Xu, J., Vijayakumar, P., Sharma, P. K., and Ghosh, U. (2022). Homomorphic encryption-based privacy-preserving federated learning in IoT-enabled healthcare system. *IEEE Transactions on Network Science and Engineering*, 10(5), 2864–2880.

Chapter 4
Splitfed learning methods for natural language processing

Amna Faisal[1], *N.Z. Jhanjhi*[1], *Sayan Kumar Ray*[1], *Gururaj H.L.*[2], *Farzeen Ashfaq*[1] *and Shampa Rani Das*[1]

The growing importance of data privacy has spurred the development of novel techniques for training natural language processing (NLP) models without compromising user confidentiality. This chapter explores two such techniques: federated learning (FL) and splitfed learning (SFL). FL enables distributed training on private datasets across devices, sharing only model updates with a central server. SFL, a sub-technique, takes a further step by splitting the model itself for training on both local devices and a central server, exchanging only intermediate results. This chapter explores SFL for privacy-preserving NLP tasks like text categorization and question answering (QA). Traditional approaches often necessitate centralized data storage, raising privacy concerns. FL and SFL address this by enabling distributed model training on user devices without sharing raw data. We discuss the benefits and shortcomings of each approach, highlighting FL's ability to handle complex models while acknowledging its potential communication overhead and performance limitations. We emphasize the relative newness of SFL and the early stages of research on its application in NLP tasks. Finally, we explore potential areas for future work, including reducing communication overhead, investigating optimal model architectures, and developing robust methods for handling non-IID data. Overcoming these challenges can ensure FL and SFL techniques have a promising future in NLP, enabling powerful model development while safeguarding user privacy.

Keywords: Deep learning; NLP; Federated learning; Split learning; Splitfed learning

4.1 Introduction

Federated learning (FL) drives progress in machine learning (ML), representing a paradigm shift in collaborative model training methodologies. Unlike conventional

[1]School of Computer Science (SCS), Taylor's University, Malaysia
[2]Department of Information Technology, Manipal Institute of Technology Bengaluru, Manipal Academy of Higher Education, India

ML approaches, FL prioritizes the primary concern of user privacy by altering the data aggregation process. In traditional ML, models are trained on centralized datasets, requiring the aggregation of vast amounts of sensitive user information in one location. However, FL diverges from this approach by distributing model training across individual devices, ensuring data remains decentralized. This decentralized framework mitigates privacy risks and empowers users with greater control over their data, marking a significant departure from conventional ML practices [1]. Figure 4.1 presents a high-level overview of FL architecture.

The unique capabilities of FL, particularly its collaborative model training approach and its focus on data privacy, have led to its diverse applications across industries. In healthcare, for instance, FL holds immense potential for tasks like patient diagnosis, disease prediction, and treatment recommendations [2]. This is particularly valuable because it allows for the development of powerful predictive models without the need to centralize sensitive patient information, ensuring compliance with data privacy regulations. Similarly, the financial services sector has embraced FL for applications such as fraud detection [3] and personalized recommendations. This enables financial institutions to leverage the power of customer data for improved services while maintaining the utmost respect for customer privacy.

FL's impact on NLP is radical. It allows participants with different goals or privacy concerns to collaborate on a shared learning program, [4] facilitating collaborative learning. Additionally, FL excels at fine-tuning pre-trained NLP models for specific tasks by leveraging vast amounts of unlabeled data, achieving impressive results [5] even with limited labeled data. This powerful combination of FL and NLP models like LSTMs, BERT, and NVFlare offers a practical solution for medical NLP tasks, enhancing patient care and clinical decision-making while ensuring data privacy and regulatory compliance. FL even empowers chatbots to learn privately from user interactions [6], with devices training locally and sharing only updates. This fosters the

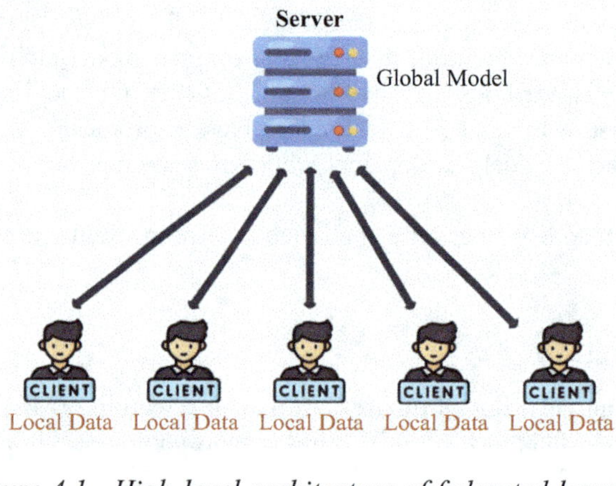

Figure 4.1 High-level architecture of federated learning

development of more effective and trustworthy chatbots. However, despite its potential, FL still faces various challenges that need to be addressed.

Due to limited processing power and battery life, FL struggles with limited resource devices needing lightweight algorithms. Uneven data distribution across devices can bias the model, and fluctuating device availability requires robust protocols for consistency. Additionally, FL suffers from longer training times due to communication overhead, raising concerns about network bandwidth and energy use in large deployments. Overcoming these challenges mandates integrative efforts and innovative solutions to optimize FL algorithms, [7] communication protocols, and resource management strategies.

On the other hand, split learning (SL) is an ML technique that addresses privacy concerns by decentralizing the model training process. It divides the model into segments [8], each stored on a different server, as shown in Figure 4.2. Local data persists on these devices, preserving privacy, and only intermediate updates to model parameters traversed between them. SL involves each participant processing local data and computing its part of the model using forward and backward passes. Computed components are shared and combined to update the global model parameters. SL has many benefits, including preserving data privacy, decreasing communication overhead due to sharing model updates instead of raw data, facilitating scalability and distributed environments, and facilitating collaboration among various parties. However, challenges such as ensuring that the model remains stable and addressing potential security risks associated with transmitting model updates continue to exist.

While focusing on privacy-preserving collaboration, SL is finding applications in different fields. Such as in healthcare, it allows wearables to train medical

Figure 4.2 High-level architecture of split learning

models collaboratively, building disease prediction tools [9] without compromising patient data. Smart cities can leverage SL for [10] traffic management, where traffic lights or sensors can collaborate on optimizing traffic flow without revealing individual data points. Moreover, even the finance industry can benefit from SL, as financial institutions can build fraud detection or risk assessment models using data from different branches without sharing sensitive customer information.

While real-world deployments are still emerging, ongoing research in NLP actively explores the potential of SF by making a way for the future of powerful and privacy-preserving NLP applications. SL offers possibilities for NLP tasks in different domains, such as federated chatbots learning from anonymized interactions or social media platforms, which analyze sentiment without revealing individual posts. Furthermore, on-device personal assistants can gain benefits from collaborative training in SL, all the while ensuring the confidentiality of text data.

SL offers advantages in speed but it's not without drawbacks. One key concern is the potential for information leakage. Even though raw data isn't shared, the intermediate results exchanged during training can reveal details about the local data used. Additionally, SL can suffer from efficiency issues. The sequential nature of training, where the server waits for updates from all clients before proceeding, can create bottlenecks. Finally, the effectiveness of SL heavily relies on the quality and balance of data distributed across different participants. Unequal data distribution can lead to biases in the final model.

SL offers faster training due to reduced client-side workload but raises privacy concerns because the server sees all model pieces. FL protects privacy better, but communication overhead can slow training. SFL offers the best of both worlds. It splits the model like SL, but instead of sharing the entire update, clients only transmit updates on their assigned portion, similar to FL. This significantly reduces communication compared to FL and offers stronger privacy guarantees than SL. Figure 4.3 provides a visual representation of the intuition behind SFL.

SFL is efficient for resource-constrained devices commonly found in privacy-sensitive applications like smart healthcare or smart cities. In healthcare systems, it can be used to train a [11] disease prediction model collaboratively. Each watch trains on its data without revealing patient information while contributing to the overall model's effectiveness. In e-commerce platforms, SFL allows the split of the recommendation model across servers based on product categories. Each server trains on its local user data and sends encrypted updates, enabling collaborative learning without revealing individual user purchase history. This improves recommendations while safeguarding user privacy.

SFL improves NLP tasks by integrating privacy preservation measures. It enables federated chatbots to collaborate on anonymized conversation data, allows social media platforms to analyze sentiment trends without compromising individual posts, and facilitates the development of multilingual language models. Beyond chatbots and sentiment analysis, SFL presents intriguing opportunities for various NLP tasks. Like text categorization, it can allow for the construction of spam filters in different organizations collaborating to improve detection. Moreover, online forums can work together to build a privacy-preserving question-answering system. SFL

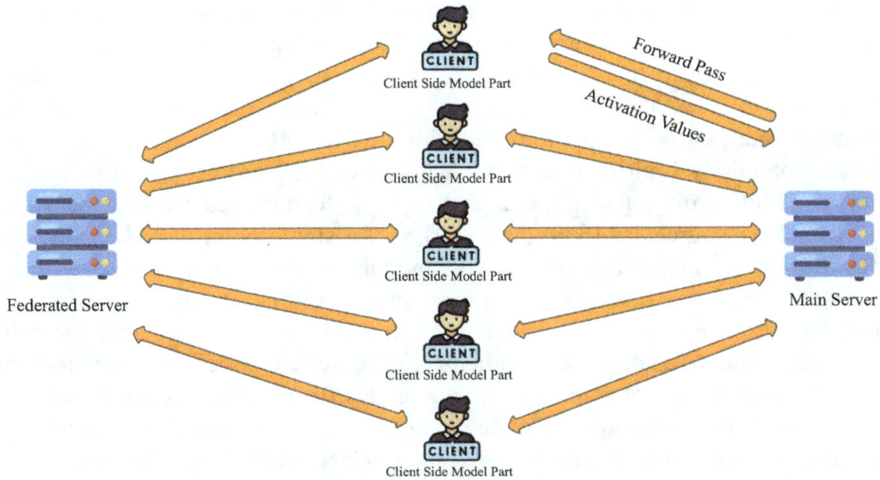

Figure 4.3 High-level architecture of splitfed learning

can even be used for on-device text analysis, such as part-of-speech tagging or federated machine translation, where devices collaboratively improve translation models without revealing the original sentences. By enabling secure collaboration, SFL opens doors for innovative and privacy-focused NLP applications. Although SFL can be applied across diverse domains of NLP, this chapter focuses specifically on its usage in text categorization and question-answering tasks.

4.2 Text categorization

Text classification, an essential aspect of natural language processing (NLP), involves classifying text documents into predefined categories based on their content. It finds application across various domains, including engineering, science, marketing, and information management systems. The main goal is to allocate each document to one or more categories, aiding in the efficient organization of textual information systems. For this purpose, various supervised ML methods like Logistic Regression, Naive Bayes, Random Forest, Support Vector Machine, and AdaBoost are frequently utilized.

Deep learning also offers powerful tools for text categorization, including convolutional neural networks (CNNs), recurrent neural networks (RNNs), and graph neural networks (GNNs). CNNs excel at identifying local patterns in text, while RNNs, especially long short-term memory (LSTM) networks, capture sequential information for [12] improved context comprehension. GNNs represent text as a network of words, [13] leveraging relationships between words for enhanced categorization. These models, in addition to others such as RNNs with attention mechanisms, aid in feature extraction and capture structural information, thereby improving the accuracy of text classification systems. While traditional ML offers a powerful approach to text categorization, it faces limitations. A major concern is [14,15] privacy, as ML often requires storing and sharing sensitive text data centrally, which can be risky for tasks like social media sentiment analysis or spam filtering. Additionally, ML relies on a single data source, potentially limiting the model's ability to handle diverse text or unseen data. Scalability may also pose a challenge, as centralized training demands considerable resources and introduces a risk of a single point of failure. In contrast, FL excels in privacy by training models on individual devices without sharing the raw text. It leverages data from various sources, leading to potentially more robust models, and its distributed training makes it more scalable and resilient. While both ML and FL are applicable, FL holds a distinct advantage for text categorization tasks that prioritize privacy and utilize various data sources. For deep learning, the large capacity of neural networks used in text classification can lead to capturing task-irrelevant patterns such as label noise [16], causing declines in [17] performance.

FL is emerging as a powerful tool for text categorization, particularly when dealing with sensitive data. Unlike traditional methods, FL keeps user privacy at the forefront by training models on individual devices without sharing the raw text itself. This is ideal for tasks like [18] spam SMS (short message service) filtering, [19] financial document classification, or [20] social media sentiment analysis, where user privacy is most important. Additionally, FL taps into the richness of data from various sources, leading to potentially more robust and generalizable models. Consider a spam filter trained on data from multiple organizations, each with unique spam patterns. The resulting model would likely outperform one trained on a single dataset. Nevertheless, the ability to leverage diverse data sources while safeguarding privacy makes FL an attractive choice for various text categorization tasks where both privacy and performance are crucial.

FL offers several promising approaches for text categorization tasks. One popular method is [21] FedAvg, which leverages the power of averaging. Here, local text classification models are trained on individual devices using their [22] local data. Crucially, the raw text data itself never leaves the devices. Instead, each device updates its local model's parameters (weights) based on the average of all participating devices' updates. This averaged update is then sent to a central server. The server aggregates these updates and uses them to improve the global model, which is then distributed back to all devices for another round of training on their local data. This iterative process allows collaborative learning without compromising privacy. While FedAvg offers a good balance between privacy and scalability for text categorization tasks, it is prone to introducing potential biases in some scenarios due to non-IID (non-independent and identically distributed) data and slower convergence.

Another approach is federated distillation. Federated distillation for text classification allows devices to train local models on their own data (emails, social media posts, etc.) [23] without sharing the raw text itself. These local models then act as "teachers," using knowledge distillation to train a smaller "student model" on the device. Only the knowledge learned by this student model is uploaded to the server, significantly reducing communication compared to sharing entire models. The server aggregates this distilled knowledge from multiple devices to improve the text classification model, making it more robust and [24] privacy-preserving. Yet, federated distillation also poses challenges with non-IID data and requires careful selection of distillation models.

Unlike FedAvg or federated distillation, which involves exchanging model parameters or knowledge between devices and a central server [25], SL keeps the training process entirely on individual devices. SL excels in privacy-preserving text categorization. Unlike other federated approaches, it keeps training on devices. The model splits: a local feature extractor analyzes local text data (emails, social media posts) and extracts informative features, like word embeddings, without revealing the raw text. These features capture the essence of categorization. A global classifier on a central server receives these features from all devices and uses them for final classification. This offers several advantages: enhanced privacy by only sharing features, reduced communication costs by transmitting features only, and potential for good scalability. For instance, hospitals can use SL to train a model for categorizing medical reports into specialties (cardiology, oncology) on their data without revealing any sensitive patient information.

FL offers a compelling approach for text classification, but it's not without its challenges. Privacy remains a concern, as even with techniques like differential privacy, there's a risk of information leakage from local data or model updates, especially for sensitive text. Another hurdle is non-IID data distribution, where text data on different devices can vary significantly. This can lead to biased models during the aggregation process in FL algorithms. Communication also poses challenges. Transferring information between devices and servers can result [26] in communication bottlenecks, especially in large-scale deployments. Techniques like model compression or selective updates can help reduce these costs. Another challenge is designing effective local models that are efficient for resource-constrained devices while maintaining accuracy. Additionally, achieving convergence (optimal model performance) in FL can be trickier compared to centralized training due to non-IID data and communication delays. In addition to these challenges, the FL model may adopt fairness and bias present in local training data, necessitating the use of techniques like federated data augmentation to address them. Security threats like poisoning attacks also pose a risk, highlighting the need for robust security measures in FL deployments.

Text categorization is revolutionized by ML, enabling tasks like spam filtering and sentiment analysis. However, traditional approaches often face significant hurdles. Centralizing vast amounts of text data on servers raises privacy concerns [21], especially when dealing with sensitive information like emails or medical records. Moreover, constraints on resources in devices such as smartphones can impede the effectiveness of conventional approaches that entail uploading extensive datasets for training. SFL emerges as an innovative solution, tackling these obstacles and opening avenues for secure and effective text categorization. Here's how it tackles these issues:

- **Privacy preservation:** Since keeping data on a centralized server raises privacy concern, SFL uses individual devices. The key difference lies in a split deep learning model. On the client side, a model resides on your device, analyzing your text data locally (like emails) to understand your unique patterns. At the same time, a model located on a central server handles more general tasks. This method effectively lowers the chance of revealing sensitive data since SFL only sends compressed information (such as text embeddings) rather than the original text.
- **Efficiency for resource-constrained devices:** SFL performs admirably in scenarios with limited device resources. By transmitting smaller data packets (compressed data), it minimizes communication overhead. This makes it ideal for devices with limited bandwidth or processing power such as smartphones in remote locations.
- **Utility and collaboration:** This approach doesn't come at the cost of sacrificing model performance. Research on SFL addresses challenges like privacy and efficiency while maintaining utility. Two prominent examples are FedSplitBERT and PPSL. FedSplitBERT tackles heterogeneous data and reduces communication costs by [21] splitting the BERT encoder layers into local and global parts. It achieves this while remaining compatible with existing FL algorithms and outperforming baseline methods, even reducing communication cost by 11.9× with quantization. PPSL, on the other hand, utilizes adversarial learning to minimize private attribute leakage in SL for [11] text classification. This demonstrates the effectiveness of SFL in preserving utility (sentiment meaning in text classification) and privacy (minimal leakage of private attributes).

With its privacy-preserving nature and efficiency gains, SFL's potential for personalization (through client-side model learning) makes it a compelling choice. Nevertheless, SFL offers a revolutionary approach to text categorization, addressing privacy concerns and resource limitations that often hinder traditional methods. With advancements in techniques, SFL demonstrates its ability to not only preserve privacy and efficiency but also [27] achieve high performance. As research continues to address challenges like optimal model splitting and compressed data security, SFL is poised to solidify its place as a powerful and secure solution for text categorization in the years to come.

4.2.1 Case study – sentiment analysis with splitfed learning for social media posts

This case study demonstrates how splitfed learning (SFL) can be applied to a text categorization task where both robust performance and strong user privacy are essential.

Scenario:

A social media platform wants to develop a system for sentiment analysis of user posts but prioritizes user privacy. They have a massive dataset of labeled posts (positive, negative, neutral) from various users. Sharing the raw post content is a privacy concern.

Solution:
SFL offers a promising approach. It leverages techniques from both SL and FL to achieve robust sentiment analysis while safeguarding user privacy.

Architecture:
Figure 4.4 presents an overview of how SFL-driven sentiment analyzers will work.

The following sequence shall be followed to implement SFL for the case study under consideration:

1. **Local feature extraction with differential privacy:** A local feature extractor analyzes the user's social media posts on each user's device. This extractor incorporates differential privacy techniques to inject noise into the features before extraction. This noise helps protect user privacy by making it difficult to infer any information about the original post content from the features alone.

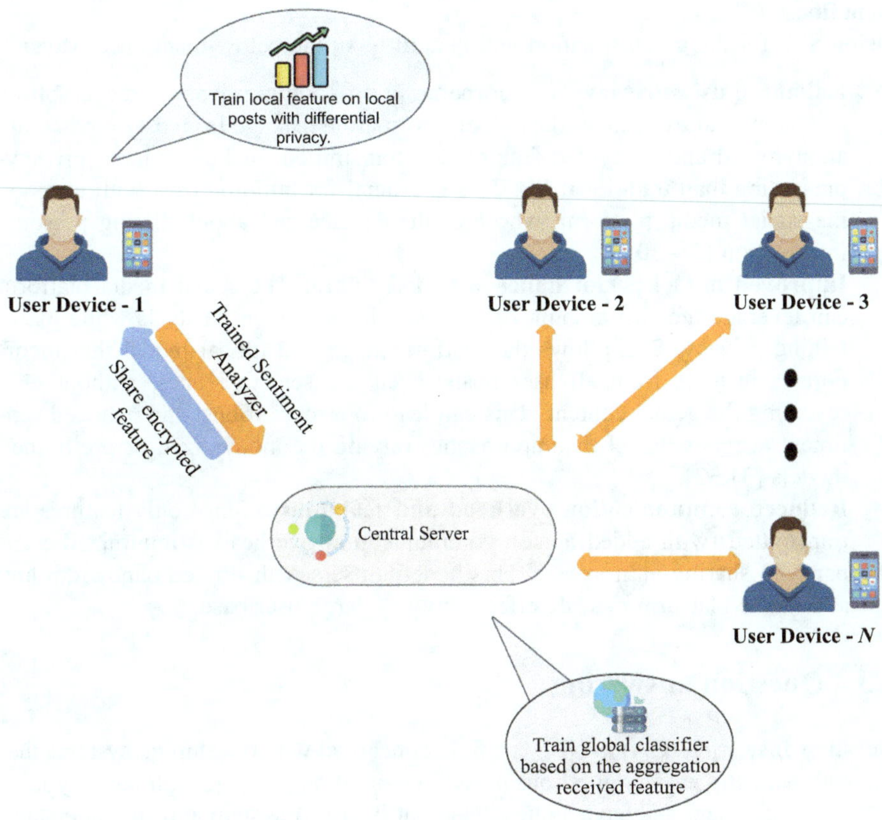

Figure 4.4 SFL-driven social media sentiment analyzer

2. **Local training:** Each device trains its local feature extractor independently on its own post data. This ensures that the raw post content never leaves the user's device.
3. **Feature sharing with encryption:** Once trained, the devices share the extracted features (with differential privacy noise added) with the central server. These features are then encrypted to provide an additional layer of security during communication.
4. **Global model training:** The central server receives the encrypted features with noise from all user devices. It then uses a secure aggregation function (potentially leveraging techniques from FL) to combine these features while minimizing information leakage. The server uses these combined features to train a global sentiment analysis model.
5. **Improved sentiment analysis:** The trained global model is then deployed back to user devices for real-time sentiment analysis. This allows the social media platform to continuously improve the accuracy of sentiment analysis by incorporating new features learned from user data over time, all while safeguarding user privacy.

Benefits:
Using SFL for the given situation will benefit us in the below-mentioned ways:

- **Enhanced user privacy:** By incorporating differential privacy and encryption, SFL significantly reduces the risk of information leakage from user posts. Only anonymized and encrypted features are transmitted, making it more privacy-preserving than traditional FL. This is crucial for building trust with users on the social media platform, who are often concerned about sharing personal information [28–30].
- **Improved model performance with user data:** The social media platform can leverage the vast amount of user data for sentiment analysis while maintaining privacy. SFL allows the platform to train a global model that incorporates insights from all user posts, even the sensitive ones, without ever revealing the actual content. This can lead to a more robust and nuanced sentiment analysis model than approaches relying on anonymized or pre-trained models [31,32].
- **Reduced communication overhead and scalability:** Since only features are transmitted (with added noise), communication overhead is minimized compared to sharing entire posts. This benefits users with limited bandwidth and allows the platform to scale effectively to a large user base.

4.3 Question answering

Question answering (QA) in NLP is a field concerned with developing systems that can automatically answer questions posed in natural language [33]. Imagine you're reading a document and have a question about it. A QA system aims to understand your question, and the document then provides a concise and informative answer, just like a helpful assistant. In the realm of NLP, QA systems aim to act like helpful assistants, understanding your questions and providing informative answers from documents. Traditional approaches to QA encompass several methods.

Information retrieval (IR) acts like a powerful search engine, sifting through documents for keywords related to the question. While efficient for factual queries [34], IR struggles with complex or open-ended questions requiring deeper context.

Extractive question answering (EQA) takes a direct approach, finding the answer within a document by pinpointing relevant sentences. This method excels at factual, short-answer questions but can struggle with complex ones requiring a deeper understanding of context.

Knowledge-based question answering (KBQA) leverages structured databases (knowledge bases) to answer questions within [35] specific domains. While powerful, KBQA faces challenges in capturing subtle connections between questions and the knowledge base, handling new information, and dealing with long-range dependencies within questions.

ML for QA utilizes algorithms like support vector machines (SVMs) to learn patterns [36] from question-answer pairs. This enables handling complex questions, but effectiveness relies on the quality and size of the training data.

These methods represent traditional approaches to question answering. As NLP advances, deep learning and neural networks are pushing the boundaries, enabling models to achieve human-like performance on complex questions and reasoning tasks.

Yet, ML models struggle to grasp the full meaning and context behind a question. They excel at providing factual answers to straightforward inquiries but falter with complex or open-ended ones requiring reasoning [37], combining information from different sources, or considering multiple interpretations. Additionally, pre-trained language models used for QA often suffer from outdated knowledge, with existing methods for updating them showing limited effectiveness [38]. Improvements are restricted to situations where the new information directly relates to existing knowledge in the model. This highlights the need for further research to enable ML models to handle the nuances of human language and keep their knowledge base current. Furthermore, the quality and quantity of data play a crucial role in the performance of ML models in QA. Insufficient data quality, limited training data, or irrelevant features may all result in inaccurate predictions [39]. Moreover, the issue of the "black box" in complex models presents challenges in comprehending their decision-making processes. Lastly, the training of extensive ML models frequently demands copious amounts of data, which raises privacy considerations. Here, FL comes to the rescue by offering a promising approach to address these issues.

FL enables the training of resilient models using distributed datasets located on user devices. This removes the necessity to centralize sensitive data and alleviates privacy worries. Furthermore, FL models can receive ongoing updates with new data from user devices, potentially addressing the challenge of outdated knowledge in pre-existing models. While challenges like communication overhead and ensuring data quality across devices remain, FL offers a compelling path toward building more robust, privacy-preserving, and adaptable question-answering systems.

FL presents an innovative method for QA in NLP, allowing for collaborative training of a robust model while protecting user privacy. FL utilizes distributed computing, allowing user devices to analyze their local documents or passages related to the question and extract valuable features, such as word embeddings or topic representations. These features capture the essence of the content without revealing the raw text. Secure aggregation techniques, like Federated Averaging, combine the extracted features from all devices on a central server. The server orchestrates the training process without directly accessing user data, thus ensuring the preservation of privacy.

This method provides several benefits for NLP QA systems. First, it prioritizes privacy by keeping user data on their devices. Second, local feature extraction on individual devices improves scalability as the central server doesn't need to handle vast amounts of raw data. Additionally, features extracted from user data can potentially capture domain-specific knowledge, leading to more accurate answers in specialized fields. Finally, FL allows for continuous learning as users interact with the system and provide new data points for the global model to be updated.

While FL for QA excels in privacy preservation, it may inadvertently leak some information through aggregated features and necessitate extensive communication for intricate models. In contrast, SL addresses these issues by transmitting smaller, processed outputs, providing enhanced privacy assurances and reducing communication overhead. Additionally, SL might enable more customized model designs for particular QA tasks compared to FL's adaptable yet potentially intricate architecture.

SL offers a compelling approach for QA tasks, particularly when user privacy is at utmost front. By keeping raw data on local devices and transmitting only processed outputs to the server, it enhances user privacy. Additionally, SL reduces communication overhead by sending smaller data chunks compared to raw text. The model architecture can also be customized, balancing privacy needs with computational demands by strategically splitting the workload. This approach can even facilitate some offline processing on user devices, increasing system flexibility.

Nonetheless, SL faces certain constraints. The sequential training process, which involves the server awaiting updates from all clients before proceeding, may result in inefficiencies. Moreover, the dependence on intermediate data storage on the server could prove cumbersome for extensive datasets. Additionally, dividing the model between the server and clients may constrain the overall complexity achievable compared to a unified, centralized model, potentially impacting performance on highly complex tasks. While SL diminishes the need for raw data transfer, it still entails regular communication between clients and the server to transmit model updates, posing a concern in scenarios with restricted bandwidth.

To address this issue, SFL frequently utilizes methods enabling local models to train simultaneously, thus easing the sequential bottleneck and markedly enhancing training speed and efficiency. This approach also enables a more adaptable architecture, where the distribution between local and server-side models can be optimized depending on the task's complexity and privacy considerations. Additionally, SFL can potentially reduce communication overhead compared to SL by strategically grouping and compressing model updates before sending them to the server. This combination of features makes SFL a powerful and scalable solution for privacy-preserving QA tasks.

4.3.1 Case study – SFL for an FAQ chatbot in a retail chain

This case study demonstrates how SFL can benefit a situation where data privacy is essential.

Scenario:
A large retail chain with multiple stores selling various products (clothing, electronics, furniture) wants to develop a centralized chatbot system. Each store has its own frequently asked questions (FAQ) section on the website, but maintaining separate FAQ sections can be cumbersome and time-consuming. To improve customer experience, the company wants to develop a chatbot system to answer common questions about products and services across all stores.

Challenges:
Below are some challenges to consider.

- Data privacy: Each store might have some location-specific FAQs (e.g., return policy variations) that they don't want to share with others.
- Data heterogeneity: The types and frequencies of questions asked can vary significantly between stores (e.g., clothing stores might get more questions about sizing, while electronics stores might get more about technical specifications).

Solution:
To develop an SFL-based chatbot system that ensures privacy preservation while enabling a high-performance, scalable, and centrally coordinated question-answering system.

Architecture:
Figure 4.5 shows the architectural overview of how SFL can be integrated in the discussed study.

The following series of steps are undertaken for the case study presented.

1. **Data collection:** Each store collects its own customer FAQs (questions and answers). These can be a mix of common and location-specific questions.
2. **Data splitting:** The questions and answers are split into two parts:
   - Public: Common questions relevant to all stores (e.g., "What are your return policy?").
   - Private: Location-specific questions relevant only to a particular store (e.g., "What are your store hours today?").

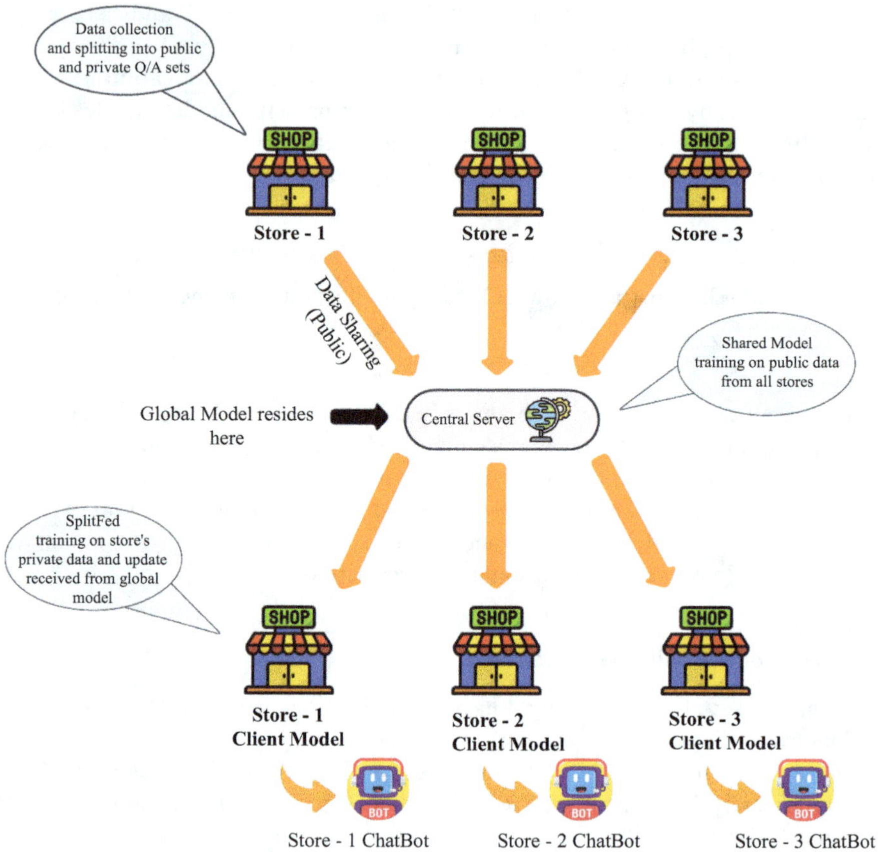

Figure 4.5 Architecture overview of SFL for QA of a retail store with multiple outlets

3. **Centralized training:** The public data is used to train a shared model (e.g., BERT) on a central server. This model captures the general knowledge about products and services.
4. **Splitfed training:** Each store keeps its private data on-site. The central server sends the shared model updates to each store's client model.
5. **Local model training:** Each store trains its local model on its private data along with the updated shared model. This allows the local model to adapt to the store's specific needs while leveraging the shared knowledge.
6. **User interaction:** Customers interact with the chatbot system through a user interface. The system identifies the store location and routes the question to the appropriate local model. The local model retrieves the answer from its combined knowledge (shared model + private data) and responds to the user.

Benefits:
A traditional QA model would typically involve aggregating data from all stores into a single central model, potentially raising privacy concerns and being both time-consuming and cumbersome. However, the proposed approach offers the following benefits:

- **Privacy preservation:** Stores only share public data for the central model, keeping location-specific information private.
- **Improved performance:** The shared model provides a strong foundation, while local models capture store-specific nuances, leading to more accurate and relevant answers.
- **Scalability:** The system can easily accommodate new stores by adding their public data to the central model and training their local models with private data.
- **Reduced centralized training load:** SFL training distributes the training workload to individual stores, reducing the computational resources needed on the central server.

4.4 Conclusion

In NLP domain, FL techniques have emerged as powerful tools for training models while preserving data privacy. FL allows distributed training on private datasets across multiple devices, with only model updates shared with a central server. This method is especially advantageous for tasks such as question answering or sentiment analysis, where protecting user privacy is of significant importance.

SL, a subset technique of FL, adopts a distinct strategy. In this method, the model is divided, with segments situated on local devices and a central server. Training then occurs on both ends, with only the intermediate results exchanged. SL offers even stronger privacy guarantees but can suffer from communication overhead and limitations on model complexity.

SFL attempts to bring closer these two approaches. It leverages the privacy benefits of SL but enables parallel training of local models, reducing training time. Moreover, SFL permits a versatile model structure, striking a balance between privacy requirements and task intricacy. Nonetheless, FL methodologies, overall, encounter obstacles such as communication overhead and potential performance limitations when contrasted with conventional centralized training methods. Nevertheless, these techniques offer a promising future for NLP tasks where data privacy and scalability are paramount. Table 4.1 draws comparison between FL and SFL.

FL, including its subfield SFL, presents exciting possibilities for NLP tasks that prioritize data privacy. However, both methods encounter challenges. FL can experience communication overhead due to frequent updates and potential limitations in performance compared to centralized training. While SFL addresses some of these issues, it remains a nascent field of research. Investigations into its optimal application for NLP tasks are still in their early stages. Future work should explore

Table 4.1 Comparison between federated learning and splitfed learning

Feature	Federated learning	Splitfed learning
Data sharing	Model updates are shared	Intermediate model outputs or gradients are shared
Privacy	Higher privacy risk (sharing model updates might reveal information)	Stronger privacy guarantees (no raw data or complete model shared)
Model architecture	Centralized model with local updates	Split model architecture (local and server-side components)
Communication overhead	Can be high depending on update frequency and data size	Potentially lower than FL due to smaller data exchange
Training efficiency	Sequential training (waits for all clients)	Can enable parallel training for faster convergence
Model complexity	Supports complex models	May limit model complexity due to split architecture
Benefits	Preserves data privacy, Scalable	Stronger privacy, reduced communication overhead (compared to FL)
Shortcomings	Communication overhead, Potential performance bottlenecks	Limited model complexity, requires careful design for efficient communication
Use cases	Question answering, sentiment analysis, text classification (when data privacy is a concern)	Similar to FL, but especially suited for scenarios with very high privacy requirements

techniques for reducing communication overhead in SFL and investigate the impact of different model architectures on NLP performance in this distributed setting. Additionally, developing robust methods for handling non-IID data, where local data distributions vary significantly across devices, is crucial for ensuring generalizability across diverse NLP tasks. As research progresses, FL and SFL hold the promise to revolutionize NLP applications by enabling robust model training while safeguarding user privacy.

References

[1] Aledhari, M., Razzak, R., Parizi, R.M. and Saeed, F., 2020. Federated learning: A survey on enabling technologies, protocols, and applications. *IEEE Access*, *8*, pp. 140699–140725.

[2] Ullah, F., Srivastava, G., Xiao, H., Ullah, S., Lin, J.C.W. and Zhao, Y., 2023. A scalable federated learning approach for collaborative smart healthcare systems with intermittent clients using medical imaging. *IEEE Journal of Biomedical and Health Informatics*.

[3] Zhang, H., Hong, J., Dong, F., Drew, S., Xue, L. and Zhou, J., 2023. A privacy-preserving hybrid federated learning framework for financial crime detection. In *KDD FL4Data-Mining '23*, ACM, Long Beach, CA, USA, pp. 1–11.

[4] Yun, W.J., Kim, S. and Kim, J., 2023. Multi-site clinical federated learning using recursive and attentive models and NVFlare. In *2023 IEEE 43rd International Conference on Distributed Computing Systems (ICDCS)* (pp. 1033–1036). Piscataway, NJ: IEEE.

[5] Dong, C., Xie, Y., Ding, B., Shen, Y. and Li, Y., 2022. Collaborating heterogeneous natural language processing tasks via federated learning. arXiv:2212.05789, https://doi.org/10.48550/arXiv.2212.05789.

[6] Ait-Mlouk, A., Alawadi, S., Toor, S. and Hellander, A., 2023. FedBot: Enhancing privacy in chatbots with federated learning. arXiv:2304.03228, https://doi.org/10.48550/arXiv.2304.03228.

[7] Dai, H. and Hong, Y., 2023. Research on model optimization technology of federated learning. In *2023 IEEE 8th International Conference on Big Data Analytics (ICBDA)* (pp. 107–112). Piscataway, NJ: IEEE.

[8] Khan, T., Nguyen, K. and Michalas, A., 2023. Split ways: Privacy-preserving training of encrypted data using split learning. arXiv:2301.08778, https://doi.org/10.48550/arXiv.2301.08778.

[9] Li, Z., Yan, C., Zhang, X., Gharibi, G., Yin, Z., Jiang, X. and Malin, B.A., 2023. Split learning for distributed collaborative training of deep learning models in health informatics. In *AMIA Annual Symposium Proceedings* (vol. 2023, p. 1047). Washington, DC: American Medical Informatics Association.

[10] Li, P., Guo, C., Xing, Y., Shi, Y., Feng, L. and Zhou, F., 2024. Core network traffic prediction based on vertical federated learning and split learning. *Scientific Reports*, *14*(1), p. 4663.

[11] Joshi, P., Thapa, C., Camtepe, S., Hasanuzzaman, M., Scully, T. and Afli, H., 2022. Performance and information leakage in splitfed learning and multi-head split learning in healthcare data and beyond. *Methods and Protocols*, *5*(4), p. 60.

[12] Kim, D., 2023. Text classification based on neural network fusion. *Tehnički glasnik*, *17*(3), pp. 359–366.

[13] Wang, Z., 2023. Deep learning based text classification methods. *Highlights in Science, Engineering and Technology*, *34*, pp. 238–243.

[14] Alnasser, W., Beigi, G., Mosallanezhad, A. and Liu, H., 2022. PPSL: Privacy-preserving text classification for split learning. In *2022 4th International Conference on Data Intelligence and Security (ICDIS)* (pp. 160–167). Piscataway, NJ: IEEE.

[15] Haralabopoulos, G. and Anagnostopoulos, I., 2022. Privacy-aware crowd labelling for machine learning tasks. arXiv:2203.01373, https://doi.org/10.48550/arXiv.2203.01373.

[16] Yang, L., Song, Y., Ren, X., *et al.* 2023. Out-of-distribution generalization in text classification: Past, present, and future. arXiv:2305.14104, https://doi.org/10.48550/arXiv.2305.14104.

[17] Gao, X., Zhu, Z., Chu, X., Wang, Y., Ruan, W. and Zhao, J., 2022. Enhancing robust text classification via category description. In *2022 IEEE International Conference on Data Mining (ICDM)* (pp. 151–160). Piscataway, NJ: IEEE.

[18] Sriraman, S., Kannan, S.S., Ravishankar, S. and Bharathi, B., 2022. An on-device federated learning system for SMS spam classification. In *2022 IEEE MIT Undergraduate Research Technology Conference (URTC)* (pp. 1–5). Piscataway, NJ: IEEE.
[19] Basu, P., Roy, T.S., Naidu, R. and Muftuoglu, Z., 2021. Privacy enabled financial text classification using differential privacy and federated learning. arXiv:2110.01643, https://doi.org/10.48550/arXiv.2110.01643.
[20] Kumar, V.N., Sivaji, U., Kanishka, G., *et al.* 2023. A framework for tweet classification and analysis on social media platform using federated learning. *Malaysian Journal of Computer Science*, Special Issue on Computing, Communication and Cyber Physical Systems, pp. 90–98.
[21] Lit, Z., Sit, S., Wang, J. and Xiao, J., 2022. Federated split BERT for heterogeneous text classification. In *2022 International Joint Conference on Neural Networks (IJCNN)* (pp. 1–8). Piscataway, NJ: IEEE.
[22] Zhang, Z., Hu, X., Qu, L., Wang, Q. and Xu, Z., 2022. Federated model decomposition with private vocabulary for text classification. Goldberg, Y., Kozareva, Z., and Zhang, Y. (eds.), In *Empirical Methods in Natural Language Processing 2022* (pp. 6413–6425). Stroudsburg, PA: Association for Computational Linguistics (ACL).
[23] Lei, Z., Yu, Y., Shi, C., Ma, Y. and Liu, J., 2024. Text-free federated transformers knowledge distillation without GAN. In *The Twelfth International Conference on Learning Representations (ICLR)*, pp. 1–14.
[24] Gong, X., Sharma, A., Karanam, S., Wu, Z., Chen, T., Doermann, D. and Innanje, A., 2022. Preserving privacy in federated learning with ensemble cross-domain knowledge distillation. In *Proceedings of the AAAI Conference on Artificial Intelligence* (vol. 36, no. 11, pp. 11891–11899).
[25] Basu, Piyu. A gentle introduction to split learning. Medium, February 17, 2021. https://piyubasu.medium.com/split-learning-vs-federated-learning-and-their-use-cases-fe96d5c1394a [accessed 9 August 2024].
[26] Huo, Z., Fan, Y. and Huang, Y., 2023. A communication-efficient federated text classification method based on parameter pruning. *Mathematics*, *11*(13), p. 2804.
[27] Chamikara, M.A.P., Bertok, P., Khalil, I., Liu, D. and Camtepe, S., 2021. Privacy preserving distributed machine learning with federated learning. *Computer Communications*, 171, 112–125.
[28] Singhal, V., Jain, S. S., Anand, D., *et al.* 2020. Artificial intelligence enabled road vehicle-train collision risk assessment framework for unmanned railway level crossings. *IEEE Access*, *8*, 113790–113806.
[29] Jhanjhi, N. Z., Brohi, S. N., Malik, N. A. and Humayun, M. 2020. Proposing a hybrid RPL protocol for rank and wormhole attack mitigation using machine learning. In *2020 2nd International Conference on Computer and Information Sciences (ICCIS)* (pp. 1–6). Piscataway, NJ: IEEE.
[30] Gaur, L., Afaq, A., Solanki, A., *et al.* 2021. Capitalizing on big data and revolutionary 5G technology: Extracting and visualizing ratings and reviews of global chain hotels. *Computers and Electrical Engineering*, *95*, 107374.

[31] Gaur, L., Singh, G., Solanki, A., *et al.* 2021. Disposition of youth in predicting sustainable development goals using the neuro-fuzzy and random forest algorithms. *Human-Centric Computing and Information Sciences, 11*, Article 24.

[32] Adeyemo, V. E., Abdullah, A., JhanJhi, N. Z., Supramaniam, M. and Balogun, A. O. 2019. Ensemble and deep-learning methods for two-class and multi-attack anomaly intrusion detection: An empirical study. *International Journal of Advanced Computer Science and Applications, 10*(9), 520–529.

[33] Pandey, A. and Bhat, A., 2023. A review on textual question answering with information retrieval and deep learning aspect. In *2023 7th International Conference on Intelligent Computing and Control Systems (ICICCS)* (pp. 224–229). Piscataway, NJ: IEEE.

[34] Tao, D., 2020. How deep learning works for information retrieval. In *Proceedings of the 43rd International ACM SIGIR Conference on Research and Development in Information Retrieval* (p. 5).

[35] Ma, L., Zhang, P., Luo, D., Zhu, X., Zhou, M., Liang, Q. and Wang, B., 2022. Syntax-based graph matching for knowledge base question answering. In *ICASSP 2022–2022 IEEE International Conference on Acoustics, Speech and Signal Processing (ICASSP)* (pp. 8227–8231). Piscataway, NJ: IEEE.

[36] Shu, X., Yan, S., Yang, X., Wu, Z., Chen, Z. and Lu, Z., 2023. SC-ML: Self-supervised counterfactual metric learning for debiased visual question answering. arXiv:2304.01647, https://doi.org/10.48550/arXiv.2304.01647.

[37] Papakostas, K. and Papadopoulou, I., 2023. Model analysis and evaluation for ambiguous question answering. arXiv:2305.12483, https://doi.org/10.48550/arXiv.2305.12483.

[38] Onoe, Y., Zhang, M.J., Padmanabhan, S., Durrett, G. and Choi, E., 2023. Can LMs learn new entities from descriptions? Challenges in propagating injected knowledge. arXiv:2305.01651, https://doi.org/10.48550/arXiv.2305.01651.

[39] Chai, C., Wang, J., Luo, Y., Niu, Z. and Li, G., 2022. Data management for machine learning: A survey. *IEEE Transactions on Knowledge and Data Engineering, 35*(5), 4646–4667.

Chapter 5

The role of splitfed learning in recommendation systems

Sujatha Krishna[1], *Badria Sulaiman Alfurhood*[2], *Amalraj Irudayasamy*[3] *and Rajesh Natarajan*[1]

A split federated learning (SFL) is an enhanced system that integrates organizer systems with united learning to address privacy concerns and provide tailored user experiences in healthcare services SFL conserves healthcare user data in the surrounding area on the client machine during model training sessions. Using the MNIST dataset, research suggested SFL with improved genetic algorithm (SFL-IGA), a paradigm for significant item commendation that protects privacy. By dividing the instruction process among local devices and a central server, the framework enables interactive model training without requiring the discovery of private user information. It's possible to improve privacy, scalability, and model flexibility in a variety of suggestion domains is highlighted by the study's exploration of its architecture and essential components. The SFL-IGA approach is shown to be more successful than other existing methods based on accuracy, precision, and F-measure features. Furthermore, the suggested model achieves performance that is equivalent to other innovative federated recommendation algorithms.

Keywords: Machine learning; Recommendation systems; Split federated learning with improved genetic algorithm (SFL-IGA); Decentralized model training; Privacy constraints; Healthcare

5.1 Introduction

A component of machine learning (ML) referred to federated learning (FL), or cooperative learning, focuses on scenarios in which numerous entities,

[1]Information Technology Department, College of Computing and Information Sciences, University of Technology and Applied Sciences-Shinas, Oman
[2]Department of Computer Sciences, College of Computer and Information Sciences, Princess Nourah bint Abdulrahman University, Saudi Arabia
[3]Information Technology Department, College of Computing and Information Sciences, University of Technology and Applied Sciences-Nizwa, Oman

including clients, interact to preserve the data decentralized while training models.

FL: Model updates are collective with an essential attendant in FL, a popular decentralized ML system where multiple users collaborate to effectively train an ML model [1]. FL has two negative effects:

- The client requires training across the entire model, which is inefficient for devices with limited resources.
- The server has complete access to client model changes.

Split learning: In contrast, split learning (SL) is an alternative decentralized approach in which server and client share the trained model. It is more beneficial for technology with limited resources since it lessens the processing load on the client device [2]. The virtual world is enhanced by design since the attendant and client have contact to both the models' components alongside each other. The clients of SL are unable to perform computations in parallel, which is an obstacle [3]. When there is just one client and the server is communicating, the other clients must wait their turn.

Splitfed: The techniques described above are combined in an innovative approach of remote collaborative ML called splitfed. It minimizes the defects of both perspectives while embracing their benefits. Comparable to SL, the concept divides computing among the server and client, enabling gadgets with limited resources to function [4]. Similar to FL, a separate server known as the FedServer collects the parallel computation of each client's modifications at each training process to adjust the global client component of the model. Because of these characteristics, splitfed learning may be used in edge installations [5]. Splitfed may change a component of the model that is displayed to the client device, which influence well-run current assaults could operate. In splitfed, a fault strike area would be smaller than in FL, considering the restrictions placed by complexity on the effectiveness of model poisoning attempts. Figure 5.1 depicts the overview of the splitfed recommendation concept.

Recommendation systems: By providing customers with precise, customized suggestions of goods based on their interests, recommendation systems (RS) aim to tackle the problem of excessive information while personalizing the user interface. Based on available data, an RS goal is to determine whether or not a user could consider an item valuable [6]. A rising number of e-commerce and retail companies, such as Amazon and eBay, have used these tactics in recent years. After collecting copious quantities of user data, these companies modify the RS to meet the needs of both the client and the business. Transport, healthcare, agriculture, e-commerce, and retail are few of the industries that depend on data from remote sensors [7]. The study's primary goal is to preserve safety standards and safeguard data between the client and server using the SFL-IGA approach.

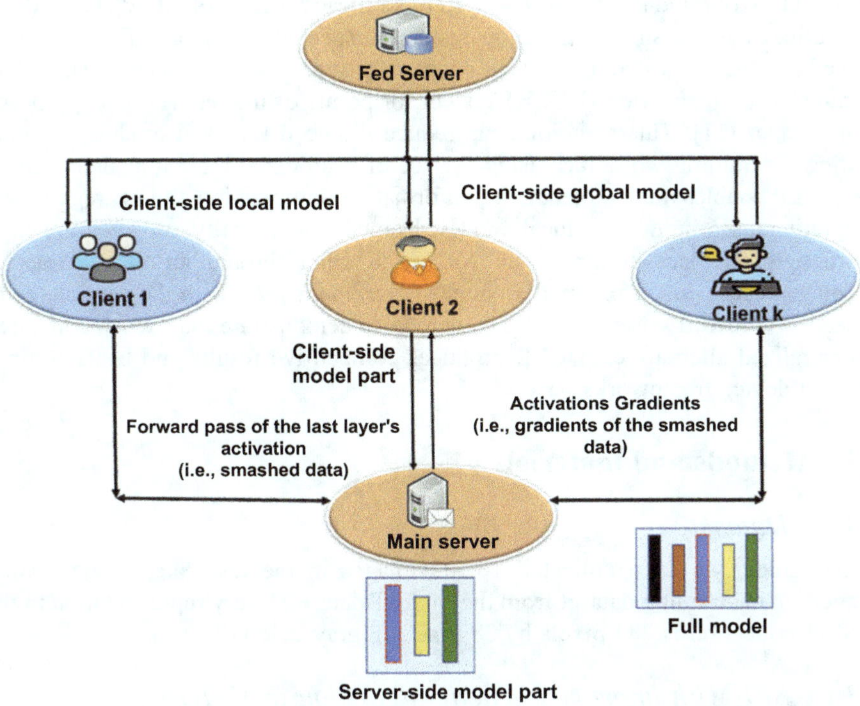

Figure 5.1 Overview of splitfed recommendation concept

5.2 Related works

An intelligent intrusion detection system (IDS) is used to protect patient medical data in IoT-based healthcare systems. It does this by utilizing edge technology, the distributed ledger, and ML [8]. The approach uses dwarf mongoose-optimized artificial neural networks (DMO-ANN) in combination with FL proposed to achieve the detection objective. This method removes contaminated assaults and offers total transparency and data integrity across the distributed network. The idea reduces processing power and computation time at the cloud's edges to defend it from attacks, and the outcomes show improved management of diverse IoT data. In this study, the integrity, availability, and confidentiality (IAC) encryption parameters are employed to protect clients and data stored in the cloud along with ownership data. Among the security techniques are the Media Access Control (MAC) and Secure Sockets Layer (SSL) protocols, searchable encryption, and sensible data partitioning. Other significant security issues covered in the article include user identity theft, unauthorized servers, and automated attacks [9]. A decision-support system for detecting deceptive instances in telecoms (TC) was industrial through a study that utilized the International Business Machines (IBM) Telco and Cell2cell datasets [10]. The technology uses cluster analysis and decision trees to classify fraud instances. The system's efficacy was testified by TC residents who utilized to pursue legal action against suspected offenders. The outcomes proved that the suggested line of action was effective. A decentralized collaborative learning recommendation (DCLR) system for points of interest (POI) is proposed in this paper [11]. This technique can reduce the need for local models on cloud hosting for training while increasing the use of centralized recommendation models. DCLR enables meticulous mutual information optimization and aggregation by integrating user input into local model deemed conceptually or geographically appropriate. Devices communicate with each other throughout this interactive learning process; just a minimal amount of communication with the central server is needed to identify user groups. The outcomes demonstrate that, when compared to centralized alternatives, DCLR produces competitive results and beats cutting-edge on-device frameworks.

5.3 Methods and materials

5.3.1 Dataset

In this study, we have collected 10,000 images in the test dataset and 60,000 imagery in the training dataset from the MNIST dataset. Every representation in the MNIST collection is 784 pixels by 28 pixels in grayscale [12].

5.3.2 SFL with improved genetic algorithm (SFL-IGA)

After collecting the healthcare data, we have a natural selection process influenced by metaheuristic, an improved genetic algorithm (IGA) uses IGA operators, such as interception, mutation, and selection, to find the best solutions from a random population. The core idea of IGA is that the fittest person would survive in a population that changes over time. When the IGA iterates over generations, crossover, mutation, and selection are the three evolutionary operators that are used. In the following generation, selection selects a group of people to merge with, while crossover joins parents to create offspring. The population's genetic variety increases because mutations can alter a people DNA at random. When a stop condition is satisfied, the IGA returns the best estimate for particular issues. Numerous operations, including population initialization, fitness score computation, crossover, mutation, and selection, are carried out by the GA. They use the initialization-of-population function to create the chromosomal population at random and the fitness-score function calculates the chromosome fitness scores. Whereas selection chooses the most suitable guardians for the following generation, crossover is the process by the best parents go on to produce children. Mutation introduces additional genetic variation to the child's chromosomes by using the mutation rate and the offspring's chromosomes as inputs. These characteristics aid in GA optimization and provide trustworthy outcomes.

A distinctive structure for a privacy-preserving recommender system based on FL, which preserves privacy presently, allows the server to create a corresponding model based on explicit responses from a portion of the user base. Equations (5.1) and (5.2) are shown and the two separate sub-networks are initialized by the Suggested Server as starting $(s = 0)$.

$$\tilde{U}_{lj}^{s} = N_u\left(U_{lj}^{s}, \theta_u^{t}\right), i \in [1, S] \tag{5.1}$$

$$\tilde{v}_l^{s} = N_v\left(V_l, \theta_v^{s}\right) \tag{5.2}$$

A cloud-edge collaboration-based SFL recommendation system. It suggests a paradigm for large-scale item suggestion that is efficient and private. We use a pragmatic cloud-edge collaborative SL technique called SFL-IGA, which stops users from disclosing their private local data but also drastically lowers the computer and communication overheads for participants with limited resources. Furthermore, we think that the SFL model that we've developed might help with the crucial computer and communication bottleneck that currently exists for FL-based recommendations. Figure 5.2 shows the architecture of the SFL system concept. SFL combines the main strengths of SL and FL, which include partitioning the network during training into server-side and client-side sub-networks and parallel processing among dispersed clients.

Clients in the simulator execute calculations concurrently with the feed server and main server. Cost-effective healthcare or Internet of Medical Things (IoM) equipment can be clients, and a cloud server with high-performance resources can

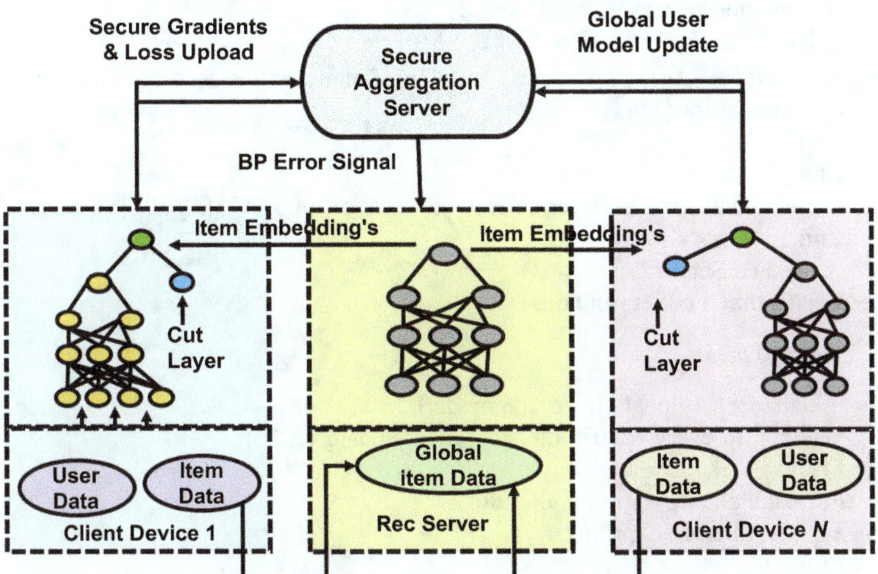

Figure 5.2 SFL architecture concept

be the primary server. In network training, the Fed Server integrates the global model and executes FedAvg on localized modifications. Based on the economical nature of the calculations, the feed server can be located inside local edge borders with SFL Algorithm 1 given below.

Algorithm 1: *SFL*

Notations: (1)$\boldsymbol{T_s}$ is a collection of t instances of L clients, (2) $\boldsymbol{B_{l,s}}$ is the corrupted client data at (3) $\boldsymbol{Z_l}$ and $\widehat{\boldsymbol{Z}}_l$ are the client's real and anticipated labels, respectively l (4) $\nabla\iota_l$ represents the client's loss gradient l (5) m and ml are the sample size at a client and the overall sample size and l, respectively.

$/ * RunsonMainServer * /$

Make sure the Main Server runs:
If, for example, time $s = 0$ **then**
Set up initially X_s^T(worldwide server-side architecture)
Else
For each client$\in$ $\boldsymbol{T_s}$ in parallel **do**
While local epoch $\boldsymbol{f \neq F do}$
$(\boldsymbol{B_{l,s}}, \boldsymbol{Z_l})$←ClientUpdate $(X_{l,s}^D)$
Propagation forward using $B_{l,s}$ *on* X_s^T, compute $\widehat{\boldsymbol{Z}}_l$
computation of loss using $\boldsymbol{Z_l}$ *and* $\widehat{\boldsymbol{Z}}_l$
Calculate backpropagation $\Delta\iota_l\ (X_s^T;\ B_s^T)$
Send $\boldsymbol{cB_{l,s}} := \Delta\iota_l(B_s^T; X_s^T)$ (i.e., gradient found in the $B_{l,s}$) to client l for ClientBackprop$(\boldsymbol{cB_{l,s}})$
End
End
model update on the server side: $X_{s+1}^T \leftarrow X_s^T - \eta\frac{m_l}{m}\sum_{j=1}^{L}\nabla\iota_l(X_s^T;B_s^T)$
End
/* use Fed Server */
Verify that FedServer runs:

ifs $= 0$ *then*

Initialize X_s^D (global client-side model)
Send X_s^D to every K customer for ClientUpdate $\left(X_{l,}^{SD}\right)$
Else
For each client$\in$ T_s in parallel **do**
$X_{l,s}^D \leftarrow$ ClientBackprop$(\mathrm{c}B_{l,s})$
End
Client-side modifications to the global model: $X_{s+1}^T \leftarrow \sum_{j=1}^{L}\frac{m_l}{m}X_{l,s}^D$
Send X_{s+1}^T to all K customers for ClientUpdate$\left(X_{l,}^{SD}\right)$
End

SFL-IGA is a decentralized approach that parallelizes genetic algorithm operations across multiple devices, allowing efficient optimization of FL models while preserving data privacy. Each device performs genetic operations on its local model fragment, guided by fitness evaluations based on healthcare data. The healthcare information is aggregated by the central server, which makes appropriate for industries including healthcare where data privacy is critical. Considering the implementation of the healthcare dataset, our suggested method performed significantly.

5.4 Experimental findings

This study evaluates the efficacy of three prediction frameworks: FedMA-SVM [13] a modified artificial bee colony radial basis function with support vector machine (MABC-RB-SVM) [13], and "an enhanced version of vanilla FedMA that combines a genetic algorithm and SVM (FedMA with GA-SVM) [13]." Based on F-measure, accuracy, and precision, it is concluded that the SFL-IGA approach is superior to other existing methods.

5.4.1 Accuracy

One of the crucial factors in calculating classification models is accuracy. In general, the proportion of forecasts is used to assess accuracy. Equation (5.3) defines accuracy and comparison and outcomes of accuracy as shown in Figure 5.3 and Table 5.1. The proposed SFL-IGA achieved (96.5%) accuracy when compared to other existing FedMA-SVM (91.7%), FedMA with GA–SVM obtained (92.3%)

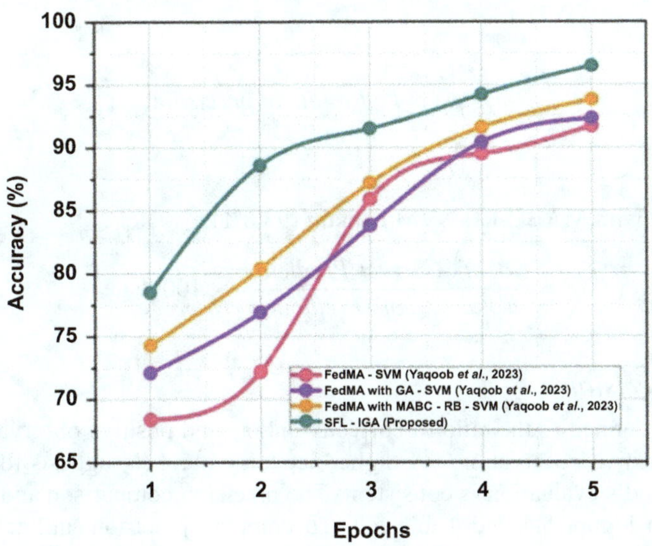

Figure 5.3 Results of accuracy

Table 5.1 Outcomes of accuracy

Epochs	Accuracy (%)			
	FedMA-SVM [13]	**FedMA with GA-SVM [13]**	**FedMA with MABC-RB-SVM [13]**	**SFL-IGA (Proposed)**
1	68.4	72.1	74.3	78.5
2	72.2	76.9	80.4	88.6
3	85.9	83.8	87.2	91.5
4	89.5	90.4	91.6	94.2
5	91.7	92.3	93.8	96.5

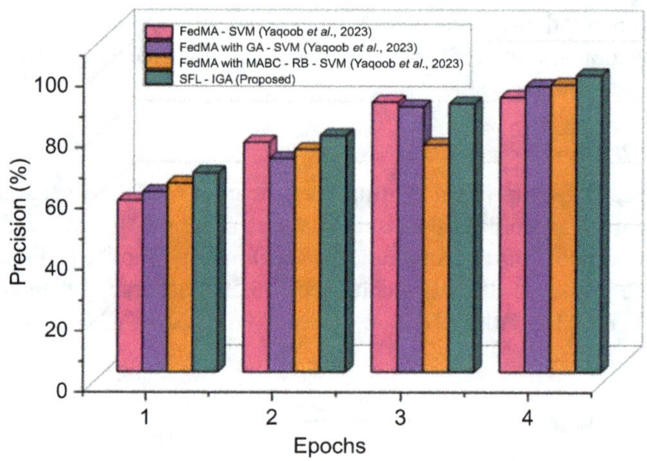

Figure 5.4 Results of precision

and FedMA with MABC-RB-SVM contain (93.8%).

$$Accuracy\ (\%) = \frac{No.\ of.\ Correct\ Prediction}{Total.\ no.\ of.\ Prediction} \times 100 \qquad (5.3)$$

5.4.2 Precision

Precision is defined as the ratio of correctly anticipated positive observations to all expected positive observations. A higher accuracy number suggests that the suggested method's evaluation is consistent. The precision comparison and results are displayed in Figure 5.4 and Table 5.2. To compute precision, utilize (5.4). The proposed SFL-IGA achieved (97.3%) precision when compared to other existing FedMA-SVM (90.1%), FedMA with GA-SVM obtained (93.7%) and FedMA with

Table 5.2 Outcomes of precision

Epochs	Precision (%)			
	FedMA-SVM [13]	**FedMA with GA-SVM [13]**	**FedMA with MABC-RB-SVM [13]**	**SFL-IGA (Proposed)**
1	56.2	59.1	62	65.3
2	75.3	69.9	72.9	77.5
3	88.5	87	74.5	88
4	90.1	93.7	94.2	97.3

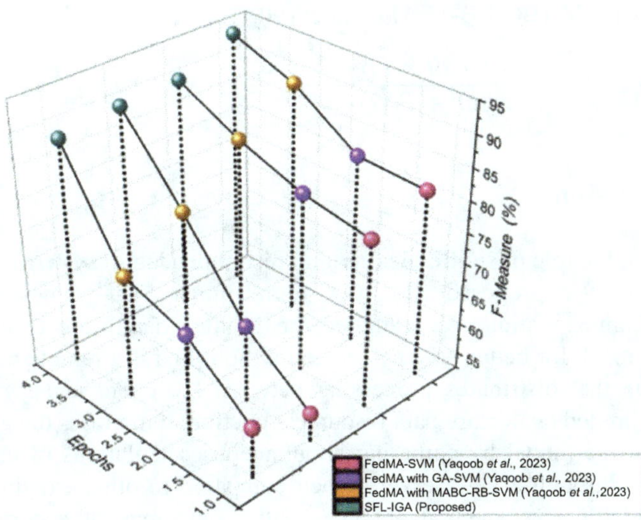

Figure 5.5 Comparison of F-measure

MABC-RB-SVM contain (94.2%).

$$Precision = \frac{TP}{TP + FP} \times 100 \tag{5.4}$$

5.4.3 F-measure

Measured as a weighted harmonic mean of the accuracy and recall, the F-measure assesses the dependability of the evaluation. The F-measure remains closer to the lower precision or recall number. Equation (5.5) provides a mathematical expression for the F-measure. Figure 5.5 and Table 5.3 show the numerical values of F-measure. The proposed SFL-IGA achieved (93.9%) accuracy when compared to other existing FedMA-SVM (88.4%), FedMA with GA–SVM obtained (89.6%)

Table 5.3 Numerical values of F-measure

Epochs	**F-measure (%)**			
	FedMA-SVM [13]	**FedMA with GA-SVM [13]**	**FedMA with MABC-RB-SVM [13]**	**SFL-IGA (Proposed)**
1	60.5	56.9	79.5	82.5
2	68.9	65	81.4	83.1
3	72.2	77.9	85	90.1
4	88.4	89.6	90.1	93.9

and FedMA with MABC-RB-SVM contain (90.1%).

$$F - measure = \frac{2 \times recall \times Precision}{recall + Precision} \quad (5.5)$$

5.5 Conclusion

In this chapter, the role of splitfed learning in recommendation systems is discussed and evaluated. We proposed SFL-IGA foundational SFL framework for the extensive recommendation. We gathered the MNIST healthcare dataset for our proposed approach for better privacy protection Split Fed is a remote collaborative ML technique that distributes processing between the client and server so that devices with limited healthcare data resources function. It modifies the global client component of the model by gathering simultaneous calculations of each client's improvements through a Fed Server. When compared to other existing methods, our proposed SFL-IGA method achieves better outcomes of accuracy contain (96.5%), precision contain (97.3%), and F-measure obtain (93.9%). By utilizing the mechanism of squeeze-and-excitation networks on the backbone model, this technique optimizes feature representation while lowering computing and communication costs. Additionally, the suggested model performs comparably to the most advanced federated recommendation models.

References

[1] Mammen, P. M. (2021). Federated learning: Opportunities and challenges. arXiv:2101.05428, https://doi.org/10.48550/arXiv.2101.05428.

[2] Wu, W., Li, M., Qu, K., Zhou, C., Shen, X., Zhuang, W., and Shi, W. (2023). Split learning over wireless networks: Parallel design and resource management. *IEEE Journal on Selected Areas in Communications*, *41*(4), 1051–1066.

[3] Pasquini, D., Ateniese, G., and Bernaschi, M. (2021, November). Unleashing the tiger: Inference attacks on split learning. In *Proceedings of the 2021 ACM*

SIGSAC Conference on Computer and Communications Security (pp. 2113–2129). https://doi.org/10.1145/3460120.3485259.

[4] Joshi, P., Thapa, C., Camtepe, S., Hasanuzzamana, M., Scully, T., and Afli, H. (2021). Splitfed learning without client-side synchronization: Analyzing client-side split network portion size to overall performance. arXiv:2109. 09246, https://doi.org/10.48550/arXiv.2109.09246.

[5] Thapa, C., Arachchige, P. C. M., Camtepe, S., and Sun, L. (2022, June). Splitfed: When federated learning meets split learning. In *Proceedings of the AAAI Conference on Artificial Intelligence* (vol. 36, no. 8, pp. 8485–8493). https://doi.org/10.1609/aaai.v36i8.20825.

[6] Fayyaz, Z., Ebrahimian, M., Nawara, D., Ibrahim, A., and Kashef, R. (2020). Recommendation systems: Algorithms, challenges, metrics, and business opportunities. *Applied Sciences*, *10*(21), 7748. https://doi.org/10.3390/app10217748.

[7] Tran, T. N. T., Felfernig, A., Trattner, C., and Holzinger, A. (2021). Recommender systems in the healthcare domain: State-of-the-art and research issues. *Journal of Intelligent Information Systems*, *57*(1), 171–201. https://doi.org/10.1007/s10844-020-00633-6.

[8] Tyagi, P., and Manju Bargavi, S. K. (2023). Using federated artificial intelligence system of intrusion detection for IoT healthcare system based on blockchain. *International Journal of Data Informatics and Intelligent Computing*, *2*(1), 1–10. https://doi.org/ 10.59461/ijdiic.v2i1.42.

[9] Jones, K. I., and Suchithra, R. (2023). Information security: A coordinated strategy to guarantee data security in cloud computing. *International Journal of Data Informatics and Intelligent Computing*, *2*(1), 11–31. https://doi.org/ 10.59461/ijdiic.v2i1.34.

[10] Khan, I. U., and Sowjanya Cherukupalli, N. L. (2023). Research on the analytics of traffic pumping in telecommunications via data science using a rehabilitated frog leaping algorithm. *International Journal of Data Informatics and Intelligent Computing*, *2*(1), 51–61. https://doi.org/10.59461/ ijdiic.v2i1.46.

[11] Long, J., Chen, T., Nguyen, Q. V. H., and Yin, H. (2023). Decentralized collaborative learning framework for next POI recommendation. *ACM Transactions on Information Systems*, *41*(3), 1–25.

[12] Joshi, P., Thapa, C., Camtepe, S., Hasanuzzaman, M., Scully, T., and Afli, H. (2022). Performance and information leakage in split fed learning and multi-head split learning in healthcare data and beyond. *Methods and Protocols*, *5*(4), 60. https://doi.org/10.3390/mps5040060.

[13] Yaqoob, M. M., Nazir, M., Khan, M. A., Qureshi, S., and Al-Rasheed, A. (2023). Hybrid classifier-based federated learning in health service providers for cardiovascular disease prediction. *Applied Sciences*, *13*(3), 1911.

Chapter 6

Reconfigurable intelligent surface (RIS)-inspired splitfed learning for over-the-air

H.B. Gurushankar[1], B.C. Soundarya[1], Gururaj H.L.[1] and Geetabai S. Hukkeri[2]

The scientific community is transitioning toward sixth-generation (6G) technology, with a focus on meeting the escalating requirement for secure wireless connectivity amidst the surge in wireless data traffic and the need for low-latency communication. The notion of reconfigurable intelligent surfaces (RIS) emerges as a solution to mitigate security vulnerabilities by intelligently manipulating wireless channel conditions, thus presenting a promising avenue for 6G wireless communication. Delving into the applications of RIS within the context of 6G, various configurations of RIS-assisted wireless systems are scrutinized, covering diverse scenarios, system and fading models, as well as performance metrics and objectives, in a comprehensive and methodical manner.

Keywords: 6G; RIS; Wireless communication

6.1 Introduction

Reconfigurable intelligent surfaces (RIS) are the next big thing in technology, and they could change wireless communication and networking forever. Unlike traditional stationary infrastructure, these systems create flexibility by adding artificial mirrors into wireless environments. These surfaces have many cheap and changeable parts that can work together to make electromagnetic waves stronger or weaker as needed, right on the spot [1].

The concept is inspired by the ability of matter to evolve and evolve, and refers to phenomena in meta materials and biological systems. RIS works by intelligently controlling the phase and amplitude of the reflected signals, essentially consisting of "smart mirrors" that transmit wireless messages to their intended targets [2]. Using sophisticated algorithms and machine learning techniques, RIS adapts itself

[1]Department of Information Technology, Manipal Institute of Technology Bengaluru, Manipal Academy of Higher Education, India
[2]Department of Computer Science and Engineering, Manipal Institute of Technology Bengaluru, Manipal Academy of Higher Education, India

to changing environmental conditions, user preferences, user requirements, and can customize its system accordingly.

RIS holds promise throughout a couple of domains, encompassing 5G and past, Internet of Things (IoT), indoor localization, and smart towns [3–5]. Within 5G networks, RIS can surmount propagation limitations, increase insurance to previously inaccessible zones, and bolster spectral efficiency. In IoT implementations, RIS streamlines connectivity, bolsters energy efficiency, and fortifies reliability. Additionally, in indoor settings, RIS helps correct localization, diminishes multipath fading, and augments high quality of carrier (Figure 6.1).

The preexisting cell networks are confronted with titanic hurdles because of the unparalleled necessities for ubiquitous and advanced pleasant wireless services. The 5th generation (5G) of communique systems is meant to be used for packages together with rate-centric better cellular broadband (eMBB), ultra-dependable, low latency communications (URLLC), and massive gadget type communications (mMTC). However, it is anticipated that the goals of the sixth-generation (6G) of wi-fi communique structures might be progressive and transformative, masking programs consisting of linked intelligence and statistics-driven, instantaneous, ultra-massive, and pervasive wi-fi connectivity [6]. Therefore, with a view to permit these new packages and offerings, new transmission technologies are required. RIS, also known as massive sensible surfaces (LIS) or sensible reflecting surfaces (IRS), are surfaces with an array of reflecting devices that permit the incident alerts to be reconfigured. Due to their capability to alter the wi-fi conversation environment in a proactive manner, RIS have drawn attention from researchers analyzing wireless communications with a view to cope with a lot of problems that arise in numerous varieties of wi-fi networks [7]. The benefits of RIS are outlined as follows:

1. Enhanced signal strength: RIS intelligently adjusts electromagnetic waves to boost signal strength, improving coverage and connectivity indoors and outdoors.

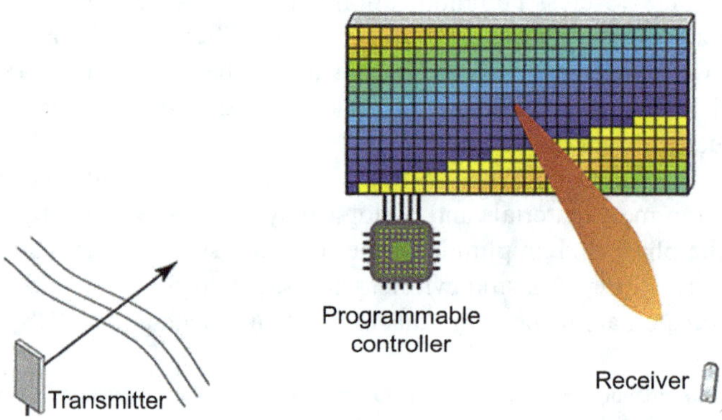

Figure 6.1 Reconfigurable intelligent surface

2. Interference mitigation: Through dynamic phase and amplitude adjustments, RIS effectively minimizes interference, enhancing wireless transmission reliability and quality.
3. Improved spectral efficiency: RIS optimizes spectrum usage by directing signals efficiently, reducing waste and meeting the growing bandwidth demand.
4. Flexibility and adaptability: Unlike fixed infrastructure, RIS adapts to environmental changes, user needs, and network dynamics, offering unparalleled flexibility in wireless deployments.
5. Cost-effectiveness: RIS comprises low-cost, reconfigurable elements, making it an economical option for enhancing wireless communication infrastructure compared to traditional methods.
6. Energy efficiency: By optimizing signal propagation and reducing high-power transmissions, RIS contributes to energy conservation, promoting environmental sustainability.
7. Enhanced security: Dynamic signal adjustments by RIS enhance wireless communication security, making it challenging for unauthorized interception or eavesdropping.
8. Indoor localization and positioning: RIS facilitates precise indoor localization and positioning by minimizing multipath fading and enhancing signal propagation, enabling applications like asset tracking and indoor navigation.
9. Support for emerging technologies: RIS enables the deployment of emerging technologies like 5G, IoT, and smart cities by addressing coverage, capacity, and reliability challenges.
10. Scalability: Highly scalable RIS solutions can be incrementally deployed to meet evolving needs, suitable for various deployment scenarios from small-scale to large-scale urban environments.

Increasing data traffic: Many humans consider that wireless conversation is a basic necessity for modern civilization. The global Internet Protocol (IP) traffic is expected to grow by 55% annually between 2020 and 2030, reaching 5,016 exabytes [1], with statistics prices accomplishing up to at least 1 Tb/s [2]. Future wi-fi networks are expected to offer a number of specific numerous services, including sensing, localization, low latency, and extremely dependable communications, further to permitting very excessive records costs. However, those desires are not covered into the architecture of 5G networks. In actuality, primary constraints emerge as necessities and desires increase. These obstacles are finally imposed with the aid of the essential characteristics of wireless functioning.

Current network design assumptions: The assumptions that the wireless environment between communicating devices (i) is maintained by nature (ii) cannot be changed, and (iii) can be compensated for by a complex system of transmission and reception that will be done solely on the specified wireless networks to be developed in the first five generations. However, after five generations of post-wireless networks, it is possible that the improvements that can be anticipated by focusing only on environmental endpoints will not be sufficient to meet the complex specifications of next-generation wireless networks and will require an architecture that integrates sensing, localization, computing, collaborative communication and guarantees very high throughput, very low latency, and very high reliability, which will need to be done with all well in real time.

An emerging paradigm—programming the environment: Breaking loose from the idea that views the wireless environment as an uncontrollable issue must bring about enormous overall performance advantages. A fashionable base station, as an instance, sends out radio waves with an energy of Watts, and user gadget choices up alerts with an electricity of μWatts. Since radio waves cannot be controlled or tailored after they may be emitted from the transmitters and earlier than they're acquired by way of the receivers, the remaining energy is normally wasted within the environment in quite a few methods, inclusive of by interfering with different network elements or posing a protection risk.

Unmanned aerial vehicles (UAVs) have become increasingly more vital in improving spectral efficiency as 5G wireless networks evolve quickly [1]. UAVs were widely used in a variety of eventualities to enhance conversation great with the aid of simultaneously optimizing UAV trajectory and communication resource allocation, along with the commonplace throughput, average secrecy price, and power efficiency maximization, amongst different matters. This is because of their high mobility, line-of-sight (LoS) transmission, and coffee cost.

RIS have received increasing attention as a way to improve resource allocation efficiency by increasing connectivity. RIS contains an abundance of economical and energy-efficient reconfigurable light sources. Each component in a RIS has the ability to exhibit the signal of a process through a controlled phase change that occurs. In order to increase the signal strength and improve the amount of availability, all phase variations can be combined to obtain a passive beam configuration, which carries signals from different transmission modes specified through matching In addition to transparent signals, the low cost of components in a RIS and the absence of transmitter modules reduce the cost of RIS implementation. The ease with which RIS can be operated in a full-duplex mode without the need for sophisticated scraping techniques is an added advantage.

6.2 Different implementations of reconfigurable intelligence surfaces

RIS can be carried out by using numerous technology and strategies, each offering particular advantages and capabilities.

Smart buildings: Within the area of clever homes, RIS offers an opportunity to optimize wireless communication and elevate indoor connectivity. Take, as an instance, a sprawling workplace complicated wherein traditional wi-fi access points come upon problems in keeping consistent coverage and high-pace connectivity because of structural limitations which include thick partitions and interference from neighboring networks. In this sort of scenario, strategically putting RIS panels embedded with reconfigurable reflecting factors throughout the constructing proves useful. These panels continually verify the wireless landscape, adjusting the segment and amplitude of reflected signals to surmount propagation boundaries and lessen interference. Moreover, RIS allows indoor localization and positioning endeavors via producing localized signal patterns, permitting specific tracking of devices and belongings in the premises.

5G networks: In the realm of 5G networks, RIS assumes a pivotal function in surmounting propagation boundaries and optimizing network performance, mainly in densely populated city settings wherein LoS propagation and signal penetration are paramount considerations. Consider a bustling urban landscape characterized through towering structures and problematic infrastructure, where traditional 5G base stations come across demanding situations in handing over reliable coverage and high throughput because of sign obstruction and multipath fading [8]. In such a state of affairs, deploying RIS on constructing exteriors, avenue furniture, and different architectural elements proves wonderful for augmenting 5G network expansions. Through dynamic modifications to surface mirrored image residences, RIS can steer and pay attention indicators closer to centered areas, assuaging signal obstruction and bolstering signal power in areas with poor coverage.

Here are two practical examples of nearly passive reconfigurable intelligent surfaces:

1. **Wireless power transfer:** Non-passive reconfigurable intelligent surfaces (NP-RIS) discover utility in wireless power transfer (WPT) structures, improving energy harvesting performance and facilitating dynamic electricity transport. In WPT situations, surfaces coated with passive resonant factors, inclusive of metamaterials or metasurfaces, seize and listen electromagnetic power from a transmitter, like a resonant coil or antenna. By integrating a confined range of lively elements, consisting of diodes or switches, NP-RIS can dynamically modify its resonant characteristics to optimize electricity seize and transfer, considering environmental elements, receiver attributes, and power desires. This capability allows green strength harvesting from ambient RF sources and helps adaptive electricity shipping for wirelessly charging digital gadgets. NP-RIS in WPT structures find software in IoT gadgets, wearables, and sensor networks, especially in which traditional stressed out energy assets are impractical or inaccessible.
2. **Environmental sensing and monitoring NP-RIS:** It can be leveraged in environmental sensing and monitoring applications to raise information series and analysis capabilities. In this state of affairs, surfaces equipped with passive sensors, together with temperature, humidity, or gas sensors, seize environmental parameters and wirelessly transmit the facts to a primary tracking system. By integrating a confined variety of active elements, such as microcontrollers or communique modules, NP-RIS helps dynamic sensor configuration and facts transmission protocols. This functionality allows the surface to adjust to evolving environmental situations, optimize sensor performance, and conserve energy by way of lowering needless facts transmission.

These examples highlight how nearly passive reconfigurable intelligent surfaces combine the benefits of both passive and active components, paving the way for innovative solutions in wireless energy transfer, environmental sensing, and beyond.

6.3 Categories of reconfigurable intelligence surfaces

RIS can be classified primarily based on various factors such as their working frequency, bodily structure, and alertness domain. Here are some commonplace forms of RIS:

1. **Passive RIS:** Passive RIS accommodates reflective elements without energetic electronic components like amplifiers or segment shifters. Instead, these surfaces utilize constant or tunable reflective materials to control electromagnetic waves. While passive RIS is commonly easier and extra strength-green than its lively counterparts, it may own limited flexibility in adjusting sign homes.
2. **Active RIS:** Active RIS integrates electronic components like segment shifters, amplifiers, and controllers to actively manage the segment and amplitude of pondered indicators. This presents more flexibility and adaptableness in controlling sign propagation, rendering energetic RIS properly acceptable for dynamic environments and superior applications necessitating actual-time changes.
3. **Analog RIS:** Analog RIS utilizes analog sign processing methods to manipulate the segment and amplitude of reflected alerts. It might also appoint segment shifters counting on varactor diodes or section modulators to obtain continuous section manage over reflected alerts. While analog RIS may provide simplicity and fee-effectiveness, it might also gift constraints in phrases of precision and scalability.
4. **Digital RIS:** Digital RIS employs virtual sign processing (DSP) strategies to precisely control the segment and amplitude of pondered alerts. It normally carries digitally managed section shifters and programmable controllers, enabling pleasant-grained adjustment of signal homes. While virtual RIS offers advanced abilities and scalability, it could be complex and highly priced compared to analog alternatives.
5. **Metasurface RIS:** Metasurface RIS makes use of metamaterials or metasurfaces with engineered electromagnetic characteristics to govern the segment, polarization, and amplitude of reflected waves. It offers benefits inclusive of compact size, low profile, and broadband operation, rendering it appropriate for packages necessitating excessive-overall performance reflective surfaces inside limited areas.
6. **Frequency-selective RIS:** Frequency-selective RIS incorporates elements that selectively control indicators in line with their frequency or wavelength. These surfaces may be engineered to mirror, transmit, or soak up precise frequency

bands, facilitating frequency-selective signal control and spectrum management in wireless communication structures.

7. **Application-specific RIS:** RIS designs can vary relying on precise utility necessities and environmental constraints. For instance, RIS custom designed for indoor wi-fi communication would possibly emphasize compact size and coffee energy consumption, whereas RIS designed for outdoor deployments may additionally prioritize robustness and weather resistance.
8. **Integrated RIS:** Integrated RIS relates to systems wherein the intelligence and manipulate mechanisms are integrated directly into present infrastructure or gadgets, consisting of partitions, windows, or IoT devices. Integrated RIS gives seamless integration with the environment and might deliver value financial savings and ease in deployment.

6.4 Fundamentals of reconfigurable intelligent surfaces

The structure and operation of a RIS may vary depending on the specific architecture and application requirements.

6.4.1 RIS structure and operation

1. **Hardware platform:**
   (a) Field-programmable gate arrays (FPGAs): FPGAs are commonly used in RIS hardware platforms due to their reconfigurability. They consist of an array of programmable logic blocks interconnected by programmable routing resources.
   (b) Other configurable hardware: Some RIS architectures may additionally utilize other configurable hardware components, including complex programmable logic devices (CPLDs) or application-specific integrated circuits (ASICs) with reconfigurable factors.
2. **Software infrastructure:**
   (a) Development tools: RIS usually employs specialized improvement gear and frameworks that facilitate the layout, synthesis, and implementation of reconfigurable hardware and software program components.
   (b) Programming models: Various programming models are used to explain the capability of the gadget, such as hardware description languages (HDLs) like Verilog or VHDL, excessive-degree synthesis (HLS) languages like C/C, and domain-particular languages (DSLs) tailored to particular utility domain names.
3. **Reconfiguration mechanisms:**
   (a) Static reconfiguration: In static reconfiguration, the configuration of the hardware stays fixed at some point of operation and is typically decided at design time or for the duration of initialization.
   (b) Dynamic reconfiguration: Dynamic reconfiguration lets in for changes within the hardware configuration in the course of runtime based on

gadget requirements or enter stimuli. This can involve reprogramming sure areas of the FPGA even as the system is operational.

4. **Configuration mode:**
   (a) During configuration mode, the FPGA masses the bitstream that defines it's logical shape and capability. This mode commonly occurs at some point of machine initialization or when transitioning between extra-ordinary running configurations.
   (b) Operation mode: In this mode, the FPGA executes the configured hardware logic to perform the desired computational tasks. It may involve parallel processing, pipelining, or other techniques to maximize performance.
5. **Control mechanisms:**
   (a) Control logic: RIS consists of control common sense responsible for coping with the reconfiguration process, coordinating facts float, and controlling the operation of the hardware and software additives.
   (b) Feedback mechanisms: Feedback mechanisms can be hired to display machine overall performance and dynamically alter the configuration or operation parameters in reaction to changing situations.
6. **Applications:**
   (a) RIS can be carried out in diverse domains such as signal processing, picture and video processing, cryptography, machine studying, and actual-time embedded structures.
   (b) Specific packages may additionally require tailored hardware configurations and software algorithms optimized for the goal undertaking or environment.

6.4.2 Frequency selective surfaces

Frequency selective surfaces (FSS) are configurations comprising elements organized periodically that showcase selectivity in the transmission, reflection, or absorption of electromagnetic waves contingent upon their frequencies. These systems find substantial usage in diverse packages related to the manipulation of electromagnetic waves, for example, in antennas, radomes, microwave filters, and stealth technology. FSS generally function normal arrays of conductive additives like patches, slots, or wires on a dielectric substrate. The periodic association and geometrical residences of those elements play an important role in figuring out the frequency reaction of the FSS. Through their capability to selectively handle electromagnetic waves based on frequency, FSS play a vital position in improving the performance of antennas by way of permitting functions such as beam shaping, polarization manage, and frequency tuning. Moreover, these surfaces can be seamlessly included into radome structures to make certain electromagnetic transparency at the same time as protective antennas from diverse environmental elements. Furthermore, FSS are considerably applied in microwave filters to regulate the transmission of alerts inside specific frequency degrees.

6.4.3 *Varactor-tuned resonators*

Varactor-tuned resonators are digital circuits or components that employ varactor diodes to adjust the resonant frequency of a resonant circuit. By adjusting the voltage applied to the varactor diodes, the effective capacitance in the resonant circuit changes, thereby altering its resonant frequency. Increasing the voltage across the varactor diodes reduces their capacitance, resulting in a higher resonant frequency, even as lowering the voltage will increase capacitance, leading to a decrease in resonant frequency. Varactor-tuned resonators normally contain a resonant circuit, which includes an LC (inductor-capacitor) tank circuit, coupled with one or more varactor diodes. Varactor diodes, also referred to as varicap diodes, are semiconductor gadgets whose capacitance varies with the applied voltage.

Varactor-tuned resonators are widely utilized in voltage controlled oscillators (VCOs), where they adjust the oscillation frequency in response to a control voltage. VCOs serve as crucial components in frequency synthesizers, phase-locked loops (PLLs), and wireless communication systems.

6.4.4 *Liquid crystal-based reflectors*

Liquid crystal-based totally reflectors make use of the specific optical characteristics of liquid crystals to control the reflection of light. Consisting of a layer of liquid crystal cloth sandwiched among two transparent electrodes—normally glass or plastic substrates—those reflectors are available in various configurations, which includes twisted nematic (TN), vertical alignment (VA), or in-aircraft switching (IPS), relying on favored optical properties. The optical traits of liquid crystals may be regulated by making use of an electric subject across the electrodes. Liquid crystals show different refractive indices based on the alignment of their molecules. By adjusting the electric subject, the orientation of liquid crystal molecules may be altered, consequently converting the refractive index and optical characteristics of the fabric. Liquid crystal-based reflectors have the potential to dynamically transition between reflective and transparent states by way of editing the carried out voltage. In the reflective nation, liquid crystal molecules align to selectively replicate incident mild of particular wavelengths, even as inside the obvious nation, molecules orient to permit mild transmission.

Liquid crystal-based reflectors are used in reflective shows, which include digital paper (e-paper) and digital shelf labels, where low power intake and excessive visibility beneath ambient mild are acceptable. Liquid crystal-based totally reflectors can be included into clever windows and architectural glazing structures to govern the transmission of sunlight and manipulate solar warmness gain. Liquid crystal-based totally reflectors are applied in optical switching gadgets, wherein they allow rapid and green routing of optical alerts in telecommunications and information verbal exchange networks. Liquid crystal primarily based reflectors require minimum energy to keep their optical kingdom, making them power-efficient for battery operated gadgets and portable applications. Liquid crystal-based totally reflectors can switch among states swiftly, allowing high-speed modulation of reflected mild.

6.5 Emerging applications of reconfigurable intelligent surfaces

RIS are constantly advancing, giving upward push to rising packages throughout numerous domain names. RIS are more and more finding deployment in side computing and IoT gadgets to facilitate real-time processing and selection-making at the network's side. Leveraging their reconfigurable nature, RIS can regulate to evolving IoT application needs, optimize aid utilization, and facilitate efficient information processing and analytics. In the area of wi-fi communique systems, RIS play a pivotal position in 5G and beyond-5G (B5G) networks by way of permitting dynamic spectrum access, beamforming, and interference mitigation techniques. They can be seamlessly incorporated into base stations and person gadget to decorate community performance, increase spectral efficiency, and facilitate new communication paradigms like big MIMO (multiple input multiple output) and mmWave communique. Figure 6.2 shows the rising packages of RIS [9].

Autonomous motors can improve safety and dependability, optimize sensor facts processing, and regulate to changing environmental occasions with the aid of integrating RIS into their sensors and processing gadgets. RIS has the ability to use organic and healthcare applications including diagnostics, clinical imaging, and customized medication. Applications for risk detection, anomaly detection, and steady conversation using RIS are receiving greater popularity in cybersecurity and defense fields. RIS are being used in energy management systems and smart grids to optimize strength distribution, track power usage, and facilitate demand-side manage. RIS are being utilized for atmosphere management, precision farming, and environmental monitoring.

6.5.1 RIS-assisted mmWave and THz communications

The exceptionally directed nature of mmWave and THz alerts makes them prone to obstructions like barriers and reflections. High direction loss, frequency-selective

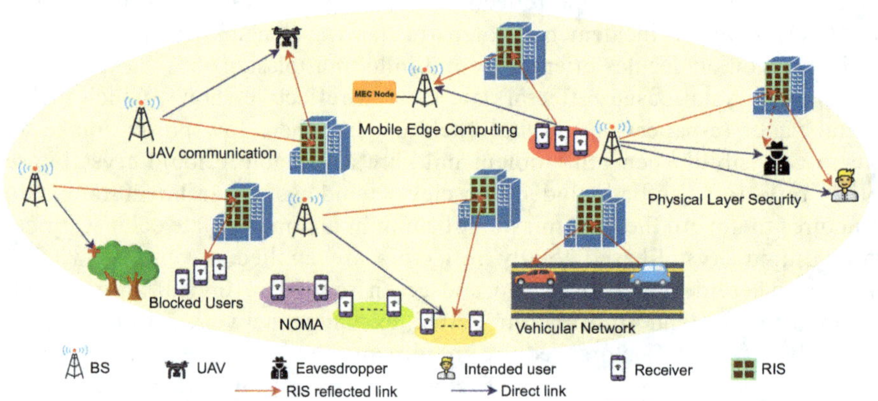

Figure 6.2 Emerging applications of reconfigurable intelligence surfaces

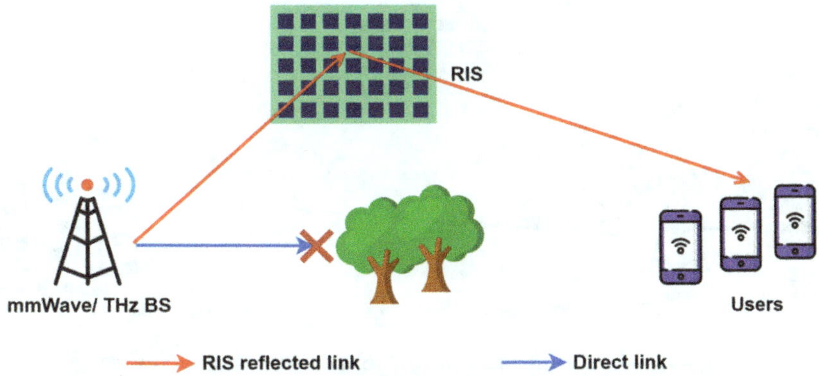

Figure 6.3 An illustration of RIS-assisted mmWave and THz communications

fading, and vulnerability to atmospheric attenuation are the traits of mmWave and THz channels. The mmWave and THz bands are liable to interference from surrounding environmental assets, system, and cells. Due to sizeable air attenuation and cloth absorption, the mmWave and THz alerts have limited coverage and penetration competencies [9]. An example of RIS-assisted mmWave and communications is offered in Figure 6.3. Though there are plenty of spectrum sources available within the mmWave and THz bands, they may be now not absolutely applied due to technological and regulatory boundaries.

6.5.2 RIS-assisted MEC networks

Computation-extensive jobs may be transferred from cell gadgets to side servers on the network aspect with the help of RIS. By processing and routing statistics intelligently at the brink, RIS can lessen the computing load on cell gadgets and beautify the efficiency of packages. In MEC networks, RIS can improve wi-fi channel situations by way of dynamically various reflection coefficients to maximize useful resource allocation and utilization. By dynamically adjusting the propagation environment to generate virtualized verbal exchange channels focused on unique programs or person companies makes community reducing and virtualization simpler in MEC networks. In dense deployment occasions, RIS-enabled beamforming algorithms enhance MEC communications reliability, ability, and spectrum performance.

RIS-enabled orchestration structures can automate provider provisioning, scaling, and optimization, permitting efficient and responsive MEC network operations. Figure 6.4 indicates the instance of RIS-assisted MEC networks [10].

6.5.3 RIS-assisted UAV networks

RIS generation holds extremely good promise for boosting the performance, coverage, and performance of UAV networks. RIS can enhance the connectivity and coverage of UAV networks through dynamically controlling the propagation

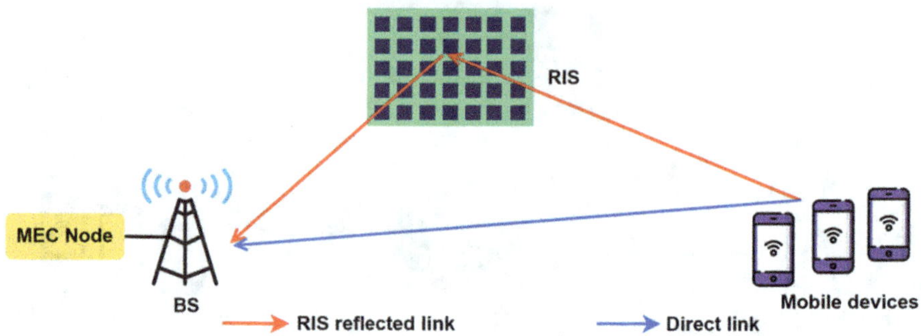

Figure 6.4 *An illustration of RIS-assisted MEC networks*

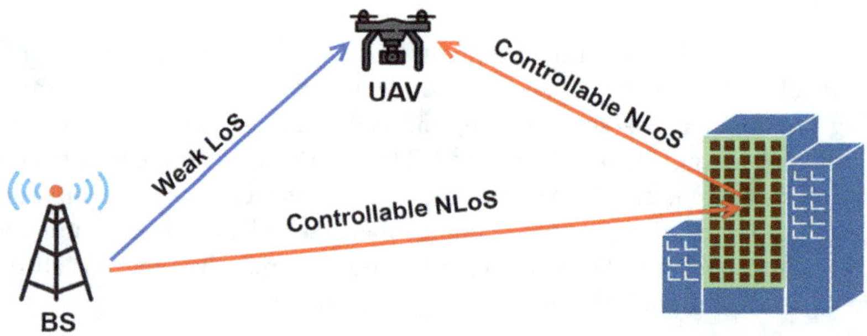

Figure 6.5 *An illustration of RIS-assisted UAV networks*

surroundings [5]. By adjusting the mirrored image coefficients of RIS factors, sign strength and satisfactory may be optimized, mitigating course loss and shadowing effects in city, rural, or harsh terrains. In congested or interference-restricted environments, RIS can help in mitigating interference and improving signal satisfactory for UAV communications. RIS-enabled beamforming strategies enable UAVs to keep strong and excessive-throughput communication hyperlinks, even in hard propagation environments or at high altitudes.

By deploying RIS elements in regions with bad insurance or excessive interference, operators can beautify the performance and reliability of UAV communications all through critical venture operations. RIS-assisted UAV networks can acquire longer venture intervals, elevated operational variety, and progressed ordinary electricity performance, permitting prolonged surveillance, tracking, and communication missions. Figure 6.5 suggests the instance of RIS-assisted UAV networks.

6.5.4 RIS-assisted VANETs

RIS holds fantastic promise for enhancing VANETs (Vehicular Ad Hoc Networks). RIS can be implemented to VANETs to enhance conversation efficiency, coverage,

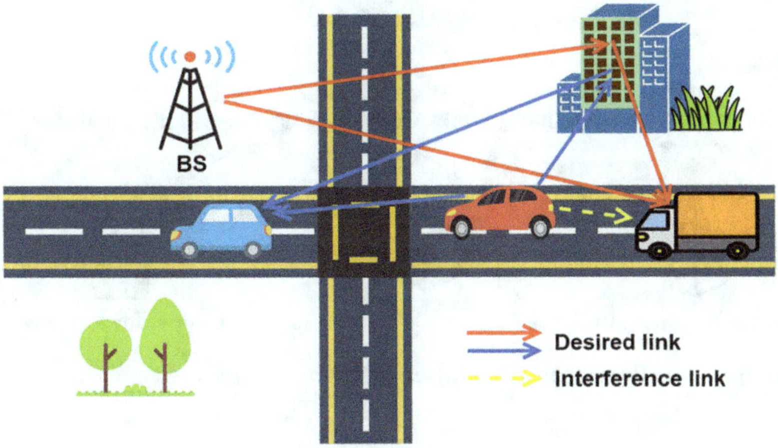

Figure 6.6 An illustration of RIS-assisted VANETS

and reliability. To enhance wi-fi communique sign coverage, RIS may be located next to roads. Through planned manipulation of the section and amplitude of pondered alerts, RIS may also successfully lessen obstructions and sign blockages, guaranteeing superior communique between cars and infrastructure. RIS is able to adjust the section and amplitude of reflected alerts in order to dynamically adapt to the propagation environment. With the use of this ability, VANET channel potential can be maximized, resulting in expanded information speeds and advanced conversation performance. VANETs may enjoy interference from adjacent vehicles and infrastructure under crowded metropolitan conditions [11].

By selectively reflecting indicators to cancel out undesired interference resources, RIS may be used to reduce such interference and increase the overall dependability of communique lines. Additionally, RIS can be used to improve the privacy and safety of VANET communications [12,13]. Ensuring the confidentiality of vital data dispatched between automobiles, RIS can forestall illegal access to communication channels and eavesdropping by way of managing the propagation surroundings. All things considered, the combination of RIS technology into VANETs holds the ability to completely rework the manner cars connect to infrastructure and one another, resulting in transportation networks which are more secure, more dependable, and extra effective. The instance of RIS-assisted VANETS communication is presented in Figure 6.6.

6.5.5 RIS-assisted IoT networks

RIS technology has the capability to significantly improve the functionality and effectiveness of IoT networks [14]. In IoT installations, RIS can be strategically positioned to increase the coverage area of wireless networks, especially in locations with obstructions and signal blockages. RIS can also efficaciously overcome obstacles and provide connectivity to IoT gadgets in remote or hard-to-attain

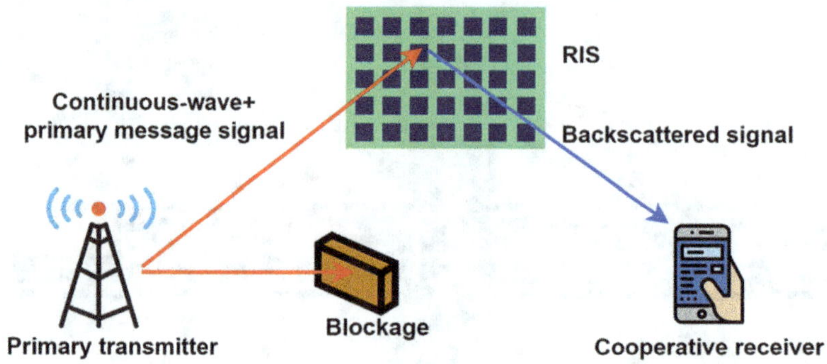

Figure 6.7 An illustration of a RIS-assisted backscatter communication system

regions by changing the phase and amplitude of meditated indicators. RIS can improve IoT network signal quality by mitigating multipath fading and signal attenuation. By optimizing the propagation environment, RIS reduces harmful interference and enhances communication link reliability, leading to stronger and more dependable connections between IoT devices and gateways.

RIS technology selectively reflects and reroutes wireless indicators to permit for dynamic spectrum management. This feature may be used to maximize using available spectrum in IoT networks, permitting several gadgets to characteristic concurrently without interfering with one another [15]. RIS-assisted IoT networks can achieve greater spectral performance and support more IoT devices within the same frequency band by efficiently allocating available spectrum resources. By maximizing the propagation surroundings to reduce wi-fi conversation's power intake, RIS can help in enhancing the energy performance of IoT devices. RIS can increase the battery life of IoT devices and permit energy-limited gadgets to run for prolonged periods of time without having normal battery replacements or recharging by minimizing the requirement for high transmission energy degrees. Figure 6.7 suggests the instance of a RIS-assisted backscatter communique machine.

6.6 Conclusion

In summary, 6G wireless communication technology has the ability to meet the growing demands of low-latency, secure, and ubiquitous connectivity. RIS are a promising solution that have the potential to revolutionize wireless communication infrastructure in the field of emerging technologies. By means of dynamic adaptation and intelligent manipulation of electromagnetic waves, RIS offers numerous benefits in a variety of application sectors. Upon careful analysis of RIS implementations, applications, classifications, and underlying concepts, it becomes evident that RIS has the significant potential to completely transform wireless networks. Through its ability to enhance signal strength, reduce interference, provide accurate localization, and support new technologies like the IoTs and smart cities, RIS proves to be a flexible solution that can meet the diverse needs of 6G wireless communication systems. In addition, the discussion of different RIS applications—from 5G networks to intelligent buildings—as well as semi-passive RIS cases like environmental sensing and wireless power transmission underscores how versatile and flexible this technology is in solving a wide range of problems.

Furthermore, classifying RIS based on several parameters such as operational frequency, physical design, and application domain provides insightful information about the diverse range of RIS solutions, each tailored to particular requirements and constraints. In the end, the fundamentals of reconfiguration methodologies, software architecture, hardware frameworks, and usage of RIS shed light on the basic principles governing the functionality and performance of systems that contain RIS. In essence, RIS integration into future wireless communication networks has the potential to reveal new features, improve efficiency, and guarantee continuous connectivity in a world that is becoming more networked by the day.

References

[1] Liu, Yuanwei, Xiao Liu, Xidong Mu, *et al.* "Reconfigurable intelligent surfaces: Principles and opportunities." *IEEE Communications Surveys & Tutorials* 23, no. 3 (2021): 1546–1577.

[2] Yuan, Xiaojun, Ying-Jun Angela Zhang, Yuanming Shi, Wenjing Yan, and Hang Liu. "Reconfigurable-intelligent-surface empowered wireless communications: Challenges and opportunities." *IEEE Wireless Communications* 28, no. 2 (2021): 136–143.

[3] Basar, Ertugrul, Marco Di Renzo, Julien De Rosny, Merouane Debbah, Mohamed-Slim Alouini, and Rui Zhang. "Wireless communications through reconfigurable intelligent surfaces." *IEEE Access* 7 (2019): 116753–116773.

[4] Björnson, Emil, Henk Wymeersch, Bho Matthiesen, Petar Popovski, Luca Sanguinetti, and Elisabeth De Carvalho. "Reconfigurable intelligent surfaces: A signal processing perspective with wireless applications." *IEEE Signal Processing Magazine* 39, no. 2 (2022): 135–158.

[5] Jian, Mengnan, George C. Alexandropoulos, Ertugrul Basar, *et al.* "Reconfigurable intelligent surfaces for wireless communications: Overview of hardware designs, channel models, and estimation techniques." *Intelligent and Converged Networks* 3, no. 1 (2022): 1–32.

[6] Björnson, Emil, and Luca Sanguinetti. "Rayleigh fading modeling and channel hardening for reconfigurable intelligent surfaces." *IEEE Wireless Communications Letters* 10, no. 4 (2020): 830–834.

[7] Di Renzo, Marco, Alessio Zappone, Merouane Debbah, *et al.* "Smart radio environments empowered by reconfigurable intelligent surfaces: How it works, state of research, and the road ahead." *IEEE Journal on Selected Areas in Communications* 38, no. 11 (2020): 2450–2525.

[8] Pan, Cunhua, Hong Ren, Kezhi Wang, *et al.* “Reconfigurable intelligent surfaces for 6G systems: Principles, applications, and research directions.” *IEEE Communications Magazine* 59, no. 6 (2021): 14–20.

[9] Kisseleff, Steven, Wallace A. Martins, Hayder Al-Hraishawi, Symeon Chatzinotas, and Björn Ottersten. “Reconfigurable intelligent surfaces for smart cities: Research challenges and opportunities.” *IEEE Open Journal of the Communications Society* 1 (2020): 1781–1797.

[10] Ai, Yun, Felipe A. P. de Figueiredo, Long Kong, Michael Cheffena, Symeon Chatzinotas, and Björn Ottersten. “Secure vehicular communications through reconfigurable intelligent surfaces.” *IEEE Transactions on Vehicular Technology* 70, no. 7 (2021): 7272–7276.

[11] Aubry, Augusto, Antonio De Maio, and Massimo Rosamilia. “Reconfigurable intelligent surfaces for N-LOS radar surveillance.” *IEEE Transactions on Vehicular Technology* 70, no. 10 (2021): 10735–10749.

[12] Huang, Chongwen, Alessio Zappone, George C. Alexandropoulos, Mérouane Debbah, and Chau Yuen. “Reconfigurable intelligent surfaces for energy efficiency in wireless communication.” *IEEE Transactions on Wireless Communications* 18, no. 8 (2019): 4157–4170.

[13] Xu, Jiaqi, Yuanwei Liu, Xidong Mu, and Octavia A. Dobre. “STAR-RISs: Simultaneous transmitting and reflecting reconfigurable intelligent surfaces.” *IEEE Communications Letters* 25, no. 9 (2021): 3134–3138.

[14] Basar, Ertugrul. “Reconfigurable intelligent surface-based index modulation: A new beyond MIMO paradigm for 6G.” *IEEE Transactions on Communications* 68, no. 5 (2020): 3187–3196.

[15] Swindlehurst, A. Lee, Gui Zhou, Rang Liu, Cunhua Pan, and Ming Li. “Channel estimation with reconfigurable intelligent surfaces—a general framework.” *Proceedings of the IEEE* 110, no. 9 (2022): 1312–1338.

Chapter 7

Enhancing computational performance in healthcare through federated learning approach

Farzeen Ashfaq[1], N.Z. Jhanjhi[1], Navid Ali Khan[1], Sayan Kumar Ray[1], Gururaj H.L.[2], Amna Faisal[1] and Shampa Rani Das[1]

With the proliferation of digital health records and medical imaging data, the healthcare sector stands at the cusp of a data-driven transformation. However, leveraging this wealth of information for computational analysis poses significant challenges, primarily concerning privacy, security, and computational performance. Federated learning has emerged as a promising solution, allowing collaborative model training across distributed healthcare institutions while preserving data privacy. This study explores the rapidly developing topic of healthcare federated learning applications with an emphasis on improving computational performance. We introduce federated learning and its relevance within healthcare contexts, particularly privacy and security challenges inherent in healthcare data sharing. Further, we explore studies involving the application of federated learning in maintaining privacy while improving model performance and assess performance benchmarks of federated learning models utilizing the healthcare datasets. Some other discussed issues include optimization techniques specific to federated learning in healthcare, such as federated averaging and differential privacy, scalability challenges and strategies for resource-efficient utilization within federated learning setups, and recent studies demonstrating the effectiveness of federated learning in various healthcare tasks purpose is to clarify metrics for performance improvement including model precision, convergence rate, resource consumption, resilience to changes in data distribution, and privacy protection. By combining numerous study findings, this chapter provides a comprehensive overview of the crucial role that federated learning plays in improving computer performance in healthcare analytics while preserving data security and privacy.

Keywords: Federated learning; Healthcare data analytics; Computational enhancement; Data privacy and security

[1]School of Computer Science (SCS), Taylor's University, Malaysia
[2]Department of Information Technology, Manipal Institute of Technology Bengaluru, Manipal Academy of Higher Education, India

7.1 Introduction

The ability of encoded algorithms to learn details from the data and perform automated functions without requiring human intervention at any stage of the process is known as artificial intelligence [1]. With advanced data analysis [2], accurate predictions and patient outcomes [3,4], and decision-making processes [5], the healthcare sector has been revolutionized by these algorithms everywhere. For example, within the field of medical imaging analysis, AI-assisted models and tools help radiologists diagnose patients accurately by identifying and categorizing abnormalities in X-rays, MRIs, and CT scans. Predictive analytics powered by AI forecast patient outcomes, facilitating early intervention and personalized treatment plans [6]. These tools have the power to accelerate drug discovery by analyzing biological data to identify potential candidates and optimize drug efficacy and safety profiles [7,8]. In the area of personalized medicinal approaches, AI-driven predictive models tailor treatments based on individual patient characteristics, while optimizing healthcare operations by streamlining processes and resource allocation [9–11].

7.1.1 The emergence of machine learning and deep learning

Machine learning (ML) and deep learning (DL) have made remarkable strides over the past few decades owing to the development in computing speed, memory, and algorithm complexity [12]. ML has paved the path for the automated analysis of data and predictive modeling through a diverse array of methodologies [13]. However, DL has revolutionized the field as it entails algorithms built on the design and function of human brain neural networks, enabling them to learn intricate patterns, representations, and abstractions from raw data [14]. These learned representations have led to unparalleled levels of performance and scalability in DL algorithms, making them ideal for complex tasks in multiple domains such as image analysis, natural language processing, and medical diagnostic [15,16].

ML has given boom to medical science, as being a powerful tool for analyzing huge amounts of patient data to improve diagnostics, treatment planning, and patient outcomes. ML algorithms can analyze complex datasets from various sources such as electronic health records (EHRs), medical imaging, and genomic data to identify patterns, predict disease risk, and personalize treatment regimens [17]. These algorithms have shown promise in assisting healthcare professionals in early disease detection, optimizing treatment strategies, and reducing medical errors [18]. With the integration of ML into medical practice, there is tremendous potential to revolutionize healthcare delivery and improve patient care [19–21].

DL, a more enhanced type of ML, has earned significant attention in the same domain due to its ability to automatically learn intricate patterns and representations from raw medical data including the CT and MRI scans, radiology reports, and tons of EHR data. In the context of diagnostics, DL algorithms achieved remarkable accuracy in various tasks such as image classification, segmentation, and object detection. With large-scale medical imaging datasets, these models can assist radiologists in detecting abnormalities, diagnosing diseases, and monitoring treatment response [22–25]. Furthermore, these automated techniques are being applied to various other areas of medical research, including drug discovery, genomics, and personalized medicine, with the potential to unlock new insights and innovations in healthcare [26–30].

7.1.2 The data for ML and DL in the context of healthcare

Data stands as an indispensable asset in the enhancement of healthcare. The field of data science, which involves the examination of large datasets [31], their careful examination, and the utilization of advanced programming, has become a crucial cornerstone for achieving advancement and success in the healthcare industry [32]. The emergence of digital health presents a promising opportunity to make accessible and personalize healthcare services; data plays a critical role in this revolutionary setting. Stakeholders can find hidden relationships and trends, as well as future projections, thanks to big data and ML [33]. However, medical data must be treated with extreme caution because it is intrinsically sensitive and contains information that may be used to identify specific individuals [34]. Figure 7.1 provides an in-depth exploration of the various components and applications of big data within the healthcare industry.

7.1.3 Data silos in medical data

The rapid advancement of ML, particularly in DL, has transformed various aspects of artificial intelligence, including tasks like image recognition and language processing. However, the effectiveness of these technologies heavily relies on access to extensive datasets [35], often referred to as big data, to train models effectively.

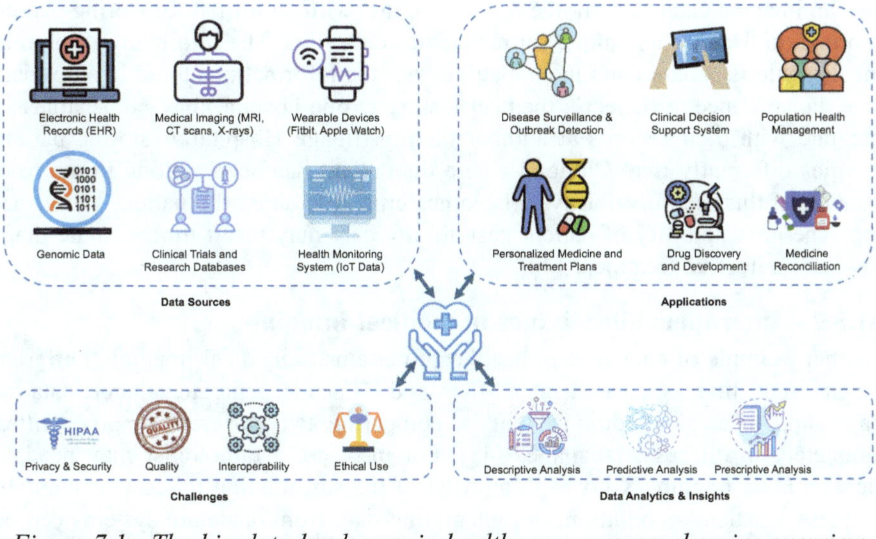

Figure 7.1 The big data landscape in healthcare: a comprehensive overview

Yet, acquiring such vast datasets can pose significant challenges [36], leading to reliance on smaller or incomplete datasets, especially prevalent in certain domains such as healthcare [37]. In medical and healthcare domains, disparate data silos are a common challenge, where patient data is fragmented across different systems and institutions [38], hindering comprehensive analysis and decision-making. In addition to this, the data here is inherently sensitive, containing information that could potentially identify patients and thus must be handled with utmost care. Patient demographics, medical histories, treatment plans, and test results are a few examples of sensitive medical data that need to be treated with extreme caution. For example, strict privacy protection is needed to avoid unauthorized access or misuse of genetic data that reveals predispositions to hereditary disorders, such as cancer or Alzheimer's [39]. Similarly, diagnostic imaging, such as Ultrasounds, MRI scans, PET scan, X-rays, containing identifiable features like facial structures or unique body marks, fluoroscopy and tomography, all demand careful handling to preserve patient anonymity. Moreover, EHRs contain a plethora of sensitive information that needs to be protected with strong security measures to avoid illegal or accidental exposure. These records, which include complete medical narratives and treatment logs, need to be carefully stored [40]. Likewise wearable health gadgets such as fitness trackers and heart monitors record physiological data in real time, which poses a privacy risk if accessed by unwanted individuals or parties. In healthcare analytics, each of these scenarios highlights the need to implement robust privacy-preserving methods to protect patient confidentiality and trust.

7.1.3.1 Electronic health records fragmentation

In the healthcare sector, one prominent challenge arises from the fragmentation of EHRs across different healthcare providers and systems [41]. The different healthcare providers include clinics and hospitals. Each uses its own EHR system. We can imagine each system like a filing cabinet with its format for storing patient information. Having its unique format creates data silos. The information stored in one data silo is isolated and incompatible with the information stored in other data silos. For instance, a patient's medical history at one hospital may not seamlessly integrate with their records at another facility. Image Hospital A storing patient allergies differently than Clinic B where the patient has been visiting in the past. Because of this incompatibility, it becomes crucial to share the patient's information. The unavailability of patient past allergy data may result in inaccurate diagnosis or ineffective treatment [42].

7.1.3.2 Interoperability issues in medical imaging

Another example of data silos in healthcare pertains to medical imaging. Different imaging modalities, such as X-rays, MRIs, and CT scans, generate valuable data for diagnosing medical conditions, that is, comparing two different reports written completely in different languages [43]. For instance, a radiologist may need to access a bone fracture X-ray with an MRI of the surrounding tissue to accurately diagnose a patient's condition, but integrating data from disparate systems can be challenging and time-consuming. This in turn limits the ability to exploit the full range of diagnostic information obtained from different imaging modalities, hence limiting the performance [44,45].

7.1.3.3 Disparate clinical trial databases

Clinical trials often involve various stages and data from multiple research institutions. Having each entity with its own data formats, protocols, and standards, the result is the formation of data silos which can also be an impediment to advancement in the fields of medication development and medical research. Clinical trial data, collected at different stages of drug testing and across various research institutions, is often stored in separate databases [46]. To identify the trends and assess treatment efficacy, the researchers need to see the whole picture together from all clinical trials at all stages across all institutes. This fragmentation makes it difficult to aggregate and analyze data comprehensively, hindering efforts to identify trends, assess treatment efficacy, and uncover potential adverse effects [47,48]. For example, adverse reactions to a drug observed in early stage trials may not be readily apparent when analyzing data from later-stage trials where the same drug was causing rare side effects, leading to potential safety risks for patients.

7.1.3.4 Privacy and ethical concerns

Unfortunately, data breaches are very common. If all the medical data is scattered across different systems, a security breach in one system can expose sensitive and confidential information [49]. This is because in a distributed environment each healthcare organization maintains its data storage system, varying security protocols, and resources. This creates multiple entry points for attackers [50]. Hence, this fragmented nature of data storage across different systems increases the risk of data breaches and unauthorized access, potentially compromising patient privacy.

7.1.4 Federated learning

In traditional ML and DL, model training typically involves aggregating a large dataset into a centralized location where the model is trained. During training, the model iteratively adjusts its parameters based on the input data to minimize the difference between predicted and actual outcomes, a process known as optimization. This centralized approach requires all data to be transferred to a single server or computing infrastructure, where computational resources are utilized to train the model. Once the model is trained, it can be deployed for inference, where it processes new data and generates predictions or classifications based on the learned patterns.

Federated learning (FL), on the other hand, introduces a decentralized approach to model training, particularly suitable for scenarios where data privacy and security are paramount. In FL, instead of transferring raw data to a central server, model training takes place locally on individual devices or edge nodes. Each device independently trains a local model using its own data while keeping the data locally stored and encrypted. After local training, only model updates, rather than raw data, are transmitted to a central server, where they are aggregated to improve

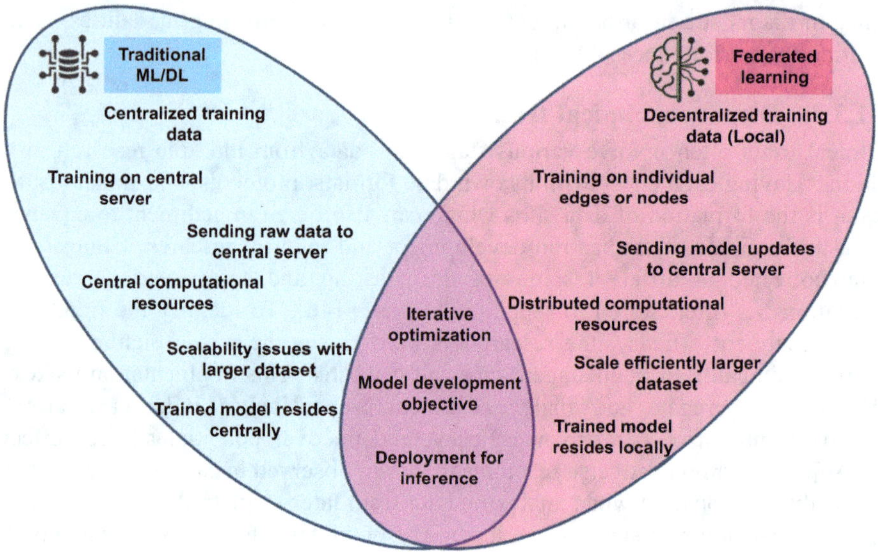

Figure 7.2 Comparison of traditional machine learning/deep learning and federated learning

the global model. This process ensures that sensitive data remains on the device, preserving user privacy while still enabling model improvement through collaborative learning [51]. Figure 7.2 provides a comprehensive comparison between FL and classical ML and DL approaches. This comparison is crucial for understanding the distinctive features, advantages, and limitations of FL in contrast to traditional centralized ML and DL methods.

7.1.5 Federated learning in healthcare and medical science

FL holds significant promise as a solution to the privacy preservation challenges inherent in healthcare data analytics while simultaneously offering computational enhancements [52]. By allowing ML models to be trained collaboratively across decentralized data sources, FL addresses concerns surrounding the privacy of sensitive medical data. With this approach, data remains localized within individual healthcare institutions or devices, mitigating the risks associated with centralized data aggregation and unauthorized access. Furthermore, FL facilitates compliance with regulations like HIPAA by ensuring that patient data remains under the control of the respective healthcare entities [53].

FL offers computational enhancements by distributing the training process across multiple devices or nodes, thereby reducing the need for data movement and central server computation. This decentralized approach not only preserves data privacy but also improves scalability and efficiency in healthcare analytics. For example, instead of transmitting large volumes of raw medical data to a central server for model training, FL allows the computation to occur locally on each device or institution's data, minimizing communication overhead and optimizing resource utilization [54].

In addition, FL enables advancements in computational performance because it leverages collective knowledge from diverse datasets without compromising data privacy [55]. For instance, in medical imaging analysis, FL algorithms can learn from a varied range of patient populations while keeping the privacy of individual medical records. This collaborative learning approach assists in the development of robust and generalizable models for tasks such as disease diagnosis and treatment planning [56–58].

7.2 Related work

In recent years, FL has gained more pull across various domains, especially in healthcare, owing to its ability to preserve data privacy and enhance model performance. It allows model training on decentralized data sources, therefore mitigating privacy risks while enabling collaboration and knowledge sharing. Figure 7.3 illustrates the taxonomy of literature covered in this review, encompassing FL's applications, advantages, and implications across different domains, with a particular focus on its transformative role in healthcare settings.

7.2.1 Need for privacy-preserving methods in healthcare data analysis

Privacy-preserving methods are key elements in healthcare data, especially in the context of big data for ensuring confidentiality, integrity, and availability in this era of AI. Between 2009 and 2023, a staggering total of 5,887 healthcare data breaches, each affecting 500 or more records, were reported to the Office for Civil Rights (OCR) [59]. This data, as compiled by The HIPAA Journal, reveals a significant trend in healthcare data security. These breaches resulted in the exposure or impermissible disclosure of an astounding 519,935,970 healthcare records, a figure exceeding the population of the United States by more than 1.5 times. The authors in [60] presents a comprehensive examination of the challenges surrounding the protection of sensitive medical research data and the facilitation of data sharing within the scientific community. Focusing on the context of the personalized Parkinson's project (PEP), conducted by Radboud University Medical Center in collaboration with Verily Life Sciences, the study sheds light on the intricate dynamics involved in balancing data privacy with the imperative of scientific progress. Through an exploration of key aspects such as informed consent, data

Figure 7.3 Literature taxonomy

Table 7.1 Reviewing health data security and privacy: related works

Application	**Need for privacy preserving**	**Related papers**
EHR security	Papers focusing on securing electronic health records systems, including authentication mechanisms, data encryption, and secure access controls to protect patient information stored in EHR databases	[61–69]
Health information exchange (HIE) security	Papers addressing security challenges and solutions in health information exchange networks, which facilitate the sharing of patient data among healthcare providers while ensuring confidentiality, integrity, and availability	[70–79]
Medical IoT security	Papers exploring security issues in medical Internet of Things (IoT) devices, such as wearable health monitors, implantable medical devices, and remote patient monitoring systems, to prevent unauthorized access, data breaches, and cyberattack	[80–89]
Telemedicine and telehealth security	Papers discussing security measures for telemedicine and telehealth platforms, including secure communication protocols, encryption techniques, and privacy-preserving data sharing mechanisms to protect patient consultations and medical data exchanged over remote channels	[90–99]
Mobile health (mHealth) security	Papers addressing security challenges in mobile health applications and devices, such as smartphone apps for health monitoring, medication adherence, and chronic disease management, to safeguard sensitive health data stored and transmitted via mobile platforms	[100–109]
Cloud-based healthcare services security	Papers examining security considerations in cloud-based healthcare services, including data encryption, access control mechanisms, and compliance, to make sure the confidentiality and integrity of patient data stored and processed in cloud environments	[110–119]

governance, data use agreements, and data security, the authors elucidate the multi-faceted considerations and pitfalls inherent in managing large-scale medical cohort studies. Table 7.1 shows the related work on health data security and privacy.

7.2.2 Privacy and security in federated learning

FL operates on a decentralized model, where model training occurs locally on individual devices or servers, and only model updates are shared with a central server. This distributed approach inherently ensures privacy and security by design. Figure 7.4 shows a basic process flow of how model updates are exchanged between the server and edge devices in FL, facilitating collaborative training while preserving data privacy and security.

FL utilizes advanced techniques to prioritize data privacy and security. For example,

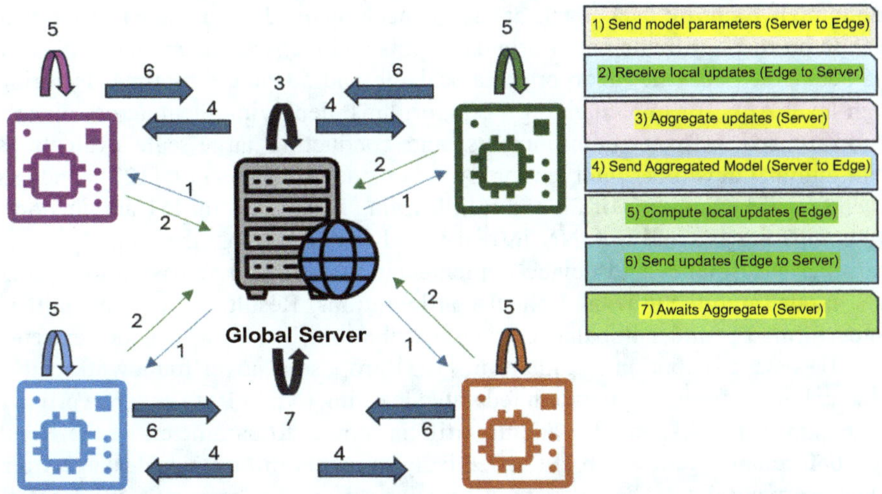

Figure 7.4 Federated learning workflow

1. Differential privacy: In this technique carefully, controlled noise is injected into the model update. In this way, the individual data points remain anonymous. Just like in our everyday scenario where to hide the actual conversation, we add unwanted background noise to the conversation, making it impossible to identify specific voices. However, the actual message is still captured and delivered.
2. Advanced encryption methods: There is one technique called homomorphic encryption that allows calculations to be performed on encrypted data. This is like working on a locked document. So even when we work the document remains locked. This method ensures that data remains encrypted even during transmissions between devices and the central server.
3. Secure aggregation protocols: These protocols ensure that the central server in the FL system only receives aggregated model updates not the individual model updates from the individual devices.

7.2.3 Performance evaluation of federated learning models

The evaluation process in FL systems involves scrutinizing various metrics such as model accuracy, convergence speed, and resource utilization across distributed devices or servers. In [120], three FL algorithms are compared against a centralized approach on the MNIST dataset, revealing that FedAvg achieves the highest accuracy across various data partitioning, highlighting its practical relevance in scenarios with heterogeneous data sources. Lai *et al.* [121] introduce FedScale, an FL benchmarking suite designed to facilitate reproducible research in FL. Such FL models offer realistic datasets covering a broad spectrum of FL tasks like categorizing an image into different disease classes, localizing tumors within an image, predicting the next drug sequence, analyzing speech patterns, or anonymizing survey data. Each dataset is accompanied by a standardized evaluation protocol utilizing real-world data splits and metrics. To emulate authentic FL behavior, FedScale incorporates a scalable and adaptable runtime, featuring high-level APIs for implementing FL algorithms, deploying them across diverse hardware and software environments, and conducting large-scale evaluations with minimal developer intervention. Another study by Gao *et al.* [122] conducts an end-to-end evaluation of FL and split learning techniques for IoT applications. By comparing FL and SplitNN in real-world IoT scenarios, the study assesses learning performance and device implementation overhead across various datasets, model architectures, and client configurations. Results show that SplitNN outperforms FL under imbalanced data distributions but lags FL under extreme non-IID data distributions. Authors in [123] propose a novel framework called noising before model aggregation federated learning (NbAFL) to enhance privacy protection in FL systems. By adding artificial noises to parameters at the client side before aggregation, NbAFL effectively prevents information leakage from clients' private data. The framework satisfies differential privacy (DP) requirements under different protection levels by adjusting the variances of the artificial noises. Table 7.2 gives the summary of performance evaluation for federated learning models in healthcare system.

Table 7.2 Summary of performance evaluation for federated learning models in healthcare system

Study	Model	Evaluation metrics	Key findings
[124]	FL with client-server architecture	AUROC, F1 curve	• FL achieved high accuracy • Performance varied with dataset distributions
[125]	Contribution-aware federated learning	Model accuracy, evaluation speed	• Significantly improves contribution evaluation speed • Enhances model accuracy by 2.62%
[126]	FL with ring signature defense	Efficiency of signature verification Success rate of SIAs	• The model effectively reduces the success rate of source inference attacks
[127]	Collaborative FL for COVID-19 screening, FL-based CNN architectures for pneumonia and cancer detection, and privacy-protected FL for skin cancer diagnosis	Performance achieved by FL models in different medical imaging tasks, such as accuracy, sensitivity, and specificity	• FL showed promising results in the detection of SARS cov2 virus using chest X-ray and CT images, with several studies focusing on collaborative FL for

(Continues)

Table 7.2 (Continued)

Study	Model	Evaluation metrics	Key findings
		Comparison of FL results with traditional machine learning (ML) models to assess the effectiveness of FL in medical image analysis	privacy-preserving screening • A significant portion of the reviewed articles (29.4%) focused on cancer detection, particularly in skin cancer diagnosis and lung cancer detection, demonstrating the potential of FL in early cancer detection
[128]	FL is employed for training models on decentralized ECG data sources across edge, fog, and cloud computing layers	• Energy consumption • Network usage • Cost • Execution time • Latency	Deployment of the FL model at the edge layer demonstrates superior performance compared to fog and cloud deployment, optimizing energy consumption, network usage, cost, execution time, and latency

7.2.4 Optimization techniques for federated learning

Several novel ideas have been put out in the field of FL optimization algorithms for healthcare IoT devices to solve the issues of accuracy and resilience under unstable network settings. Houssein and Sayed [129] introduced a novel method, termed federated improved particle swarm optimization (FedImpPSO), which enhances the FL algorithm's performance by updating the global model using score values from learned models. This approach, evaluated using CIFAR-10 and CIFAR-100 datasets, demonstrated an average accuracy improvement of 8.14% over FedAvg and 2.5% over federated PSO (FedPSO). Moreover, two case studies were conducted to evaluate FedImpPSO's effectiveness in healthcare, focusing on COVID-19 classification using ultrasound and X-ray images and cardiovascular disease prediction. The results showcased impressive F1-measure scores of 77.90% and 92.16% for COVID-19 detection and accuracies of 91.18% and 92% for predicting heart diseases, respectively. In another study [130], hybrid classifier-based FL in health service providers for cardiovascular disease prediction, an innovative hybrid framework, was proposed at the client end of health service providers (HSPs) to optimize feature selection and classification of heart disease using modified artificial bee colony optimization with support vector machine (MABC-SVM). The federated matched averaging technique was employed at the HSP server to overcome privacy concerns, resulting in improved prediction accuracy, reduced classification error, and decreased training latency compared to standard

FL techniques [131,132]. All these studies lead to the need for research in optimization techniques to enhance the performance and applicability of FL systems in healthcare.

7.3 Discussion on parameters for performance enhancement

7.3.1 Model accuracy

The use of FL techniques has shown promising results in enhancing the accuracy of predictive models because the data here remains at the individual hospitals or clinics (edge devices), compared to centralized approaches, particularly in scenarios where data heterogeneity and privacy constraints are common [133]. In the healthcare setting, traditional centralized methods often involve gathering sensitive patient data in one place for model training compromising its generalizability and accuracy. This has raised concerns about privacy breaches and regulatory compliance. A data breach at a central server exposes the data of a large number of patients [134–138].

7.3.2 Convergence speed

The speed at which FL algorithms converge (reach optimal performance) can be a major challenge and determines the overall training time. One key challenge is the data heterogeneity across distributed sources. Healthcare data from different sources may have varying distributions (skewness), which can pose challenges for efficient convergence. Other factors such as sparsity (missing data points) and outliers (unusual cases) can affect convergence speed. Addressing these challenges requires careful algorithm design and optimization techniques tailored to healthcare data characteristics.

Another challenge in the healthcare FL system is communication. Compared to the centralized system training, the data is readily available in one place; they require regular feedback between the edge devices and the central server that aggregates the learning process. Sometimes because of the network latency (delays) and limited bandwidth (data transfer capacity), this communication becomes very slow posing challenges and bottlenecks. To overcome these challenges and expedite convergence, it's imperative to optimize communication protocols using simple model design and selective model updates, to achieve faster convergence on edge devices [139,140].

Aggregation techniques that effectively combine locally trained model updates while minimizing noise and preserving privacy can lead to faster convergence [141]. It is also very important to find a tradeoff between convergence speed and model accuracy in healthcare FL [142–148]. While techniques like aggressive communication or overly simplified model updates might speed up the convergence process, they can also compromise on the model accuracy and ability to learn the intricate details in diversified medical datasets. We can compare this with a very simple example of pushing a doctor to learn quickly at the expense of fully understanding the material. The doctor might be able to get through the training program quickly, but they might not be as prepared for diagnose patient accurately.

7.3.3 Resource utilization

In health care, FL systems offer powerful approaches and techniques to train models. The focus is on understanding the resource needs of FL systems to manage the computational power, memory, and network bandwidth usage. FL systems usually entail collaboration among multiple edge devices or institutions to train a global model while maintaining decentralized and private data. In the following section, we will examine the resource utilization and optimization techniques in FL systems:

1. Computational power: Healthcare data usually consists of huge and complex medical records containing information like image scans. Processing and analyzing this large amount of data under ML requires significant computational resources on individual devices. The FL systems require computational power for not only training but also for updating models across distributed devices. One constructive approach for optimizing computational power involves model compression (both in terms of model architecture and reducing the number of model parameters); this reduces the computational overhead without compromising accuracy. Researchers are also exploring more efficient techniques involving specialized GPUs and AI chips on devices to improve overall processing speed. Another technique that greatly reduces the computational burden is selective training and federated distillation, where not all the nodes participate in all training loops. Instead, a master model is used to train smaller models on individual devices that help distribute the workload across the devices and reduce the computational demands.
2. Memory: There are three important factors that influence the memory usage in FL systems including dataset size, neural model parameters, and communication buffers. Some of the efficient memory management strategies include minimizing memory footprint through parameter quantization and sparsity techniques. In another technique based on incremental learning approach, the models are updated gradually over time instead of storing entire datasets in memory all once; this in turn can reduce memory requirements while improving the scalability issues.
3. Network bandwidth: The FL model heavily relies on network bandwidth for exchanging the model updates and gradients between edge or node devices to the central server. Network bandwidth optimization techniques focus on lowering the communication overhead while ensuring timely updates. This can be achieved through differential privacy mechanisms, where noise is added to gradients to protect sensitive information during transmission. Additionally, employing FL with model compression techniques reduces the size of transmitted data, hence conserving network bandwidth [141,149,150].

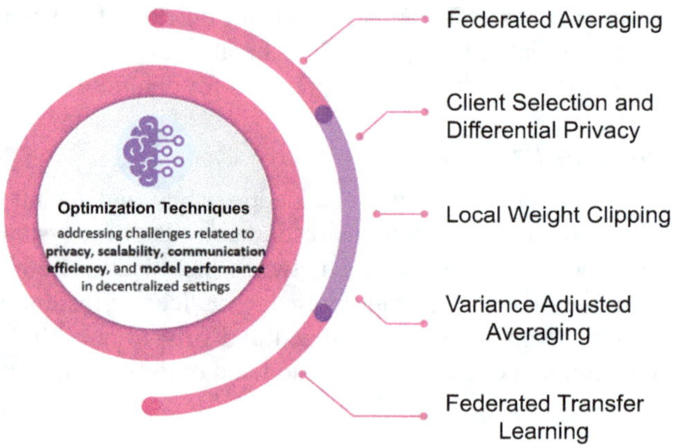

Figure 7.5 Federated learning optimization techniques applied in healthcare

4. Optimization techniques: Various optimization techniques can enhance resource utilization in FL systems. Figure 7.5 shows some of the methods so far employed in optimizing the performance of FL for healthcare data setting. For example, FL with local adaptation allows edge devices to fine-tune global models using local data [151], reducing the need for extensive communication and computation. Also, adaptive learning rate algorithms dynamically adjust learning rates based on model performance, optimizing convergence speed and resource utilization. Likewise, hierarchical FL architectures, with the introduction of intermediate aggregators to reduce communication overhead, further optimize resource utilization in large-scale FL systems. Some other techniques like local weight clipping or variance-adjusted averaging mitigate the non-identical and independent distributed data problem by reducing the influence of outliers in model updates [152,153].

7.4 Case study: cluster primal-dual splitting

To predict future hospitalizations of patients with cardiac-related diseases based on the distributed EHRs from various data sources, the performance evaluation of this research work involves several key components and optimizations [141].

1. Performance: The performance of the cPDS algorithm is evaluated using area under the receiver operating curve. The metric measures the trade-offs between true positives and false positives rate. Also, the computation time and communication costs are measured.
2. Data preprocessing: The EHR data, like setting target time intervals, selecting target years for hospitalization prediction, summarizing features over different time blocks, removing patients with no data, random splitting of the data into

training and test sets, and normalizing features. Also, the balancing of the training set by oversampling the positive class (hospitalized patients) is done.

3. Experimental evaluation: The experiments were conducted to evaluate the performance of the cPDS algorithm under different scenarios. First, the number of hospitals (agents) and graph topologies (random, cyclic, fully connected) were varied to investigate their impact on convergence. Then, comparison of the performance of cPDS with centralized methods and other distributed optimization algorithms like subgradient descent, incremental subgradient, and linear average consensus (LAC) was made.
4. Analysis of results: The experimental results show that cPDS achieves competitive AUC compared to centralized methods and other distributed algorithms while improving convergence rate. Upon the analysis of the impact of the number of hospitals and graph topologies on convergence time and communication cost, it was identified that more edges in the graph lead to faster convergence due to faster information exchange but also increase communication cost. The algorithm has pinpointed and emphasized crucial features, showcasing their significance in predicting hospitalizations for heart-related diseases.

7.5 Conclusion

In conclusion, our work addresses important issues related to data security and privacy, while highlighting the significant role that FL plays in improving computational efficiency in healthcare data. Generalizable and reproducible ML algorithms can be achieved in healthcare using FL secure methods. FL eliminates the need for data exchange or centralization by enabling multiple agents to train collaboratively. Sensitive medical data transfer concerns are addressed by FL. In most cases, FL achieves better error levels in local test data than local models on local test data by including more and more varied samples from the arbitrary discrete training dataset.

We had a considerable discussion about the challenges of sharing healthcare data, particularly in terms of security and privacy which underscores the need for more reliable techniques in FL, for example, FedAvg and FedSDG. We recommend that future researchers should include the introduction of cross-silos and cross-devices mechanisms to minimize the overall global loss and protect participants' privacy in future real-world healthcare FL projects.

Finally, we have outlined when it is possible for FL to enhance privacy without sacrificing model performance and described several criteria, including privacy attacks and distribution and partitioning. By revealing the scaling issues and outlining the most likely application strategies and FL models, the healthcare usages seem apparent, giving us an understanding of the identified applicable FL-specific fine-tuned strategies to optimize computation performance. These strategies are aggregated averages on the ML model and contrastive learning approaches to reduce heterogeneity in the dataset. Moreover, we have presented metrics for different performance improvements such as model precision, convergence rate, and communication overhead.

References

[1] Alowais, S. A., Alghamdi, S. S., Alsuhebany, N., *et al.* (2023). Revolutionizing healthcare: The role of artificial intelligence in clinical practice. *BMC Medical Education*, 23(1), 689.

[2] Shortreed, S. M., Cook, A. J., Coley, R. Y., Bobb, J. F., and Nelson, J. C. (2019). Challenges and opportunities for using big health care data to advance medical science and public health. *American Journal of Epidemiology*, 188(5), 851–861.

[3] Raghupathi, W., and Raghupathi, V. (2014). Big data analytics in healthcare: Promise and potential. *Health Information Science and Systems*, 2, 1–10.

[4] Darda, P., and Matta, N. (2024). The nexus of healthcare and technology: A thematic analysis of digital transformation through artificial intelligence. In *Transformative Approaches to Patient Literacy and Healthcare Innovation* (pp. 261–282). Hershey, PA: IGI Global.

[5] Towfek, S. K. (2023). A semantic approach for extracting the medical association rules. *Journal of Artificial Intelligence and Metaheuristics*, *5*(1), 46–52.

[6] Kasula, B. Y. (2023). Machine learning applications in diabetic healthcare: A comprehensive analysis and predictive modeling. *International Numeric Journal of Machine Learning and Robots*, 7(7), 1–9.

[7] Rana, M. S., and Shuford, J. (2024). AI in healthcare: Transforming patient care through predictive analytics and decision support systems. *Journal of Artificial Intelligence General science (JAIGS)*, 1(1), 1–4.

[8] Mistry, J. (2024). Impact of model selection on pulmonary effusion diagnosis using prediction analysis algorithms. *Journal of Xidian University*, 18 (1), 611–618.

[9] Geetha, S., and Trivedi, T. (2024). An AI-driven multi-management system for medical resource allocation providing personalized healthcare services. In Dulhare, U. N., Kumar, A. V. S., Dutta, A., Bri, S., and El Emary, I. M. M. (eds.), *Handbook of Research on Artificial Intelligence and Soft Computing Techniques in Personalized Healthcare Services* (pp. 121–136). Palm Bay, FL: Apple Academic Press.

[10] Tan, S. K., Chong, S. C., Wee, K. K., and Chong, L. Y. (2024). Personalized healthcare: A comprehensive approach for symptom diagnosis and hospital recommendations using AI and location services. *Journal of Informatics and Web Engineering*, 3(1), 117–135.

[11] Rath, K. C., Khang, A., Rath, S. K., Satapathy, N., Satapathy, S. K., and Kar, S. (2024). Artificial intelligence (AI)-enabled technology in medicine-advancing holistic healthcare monitoring and control systems. In Khang, A., Abdullayev, V., Hrybiuk, O., and Shukla, A. K. (eds.), *Computer Vision and*

AI-Integrated IoT Technologies in the Medical Ecosystem (pp. 87–108). Boca Raton, FL: CRC Press.

[12] Jiao, Y., and Du, P. (2016). Performance measures in evaluating machine learning based bioinformatics predictors for classifications. *Quantitative Biology*, 4, 320–330.

[13] Merghadi, A., Yunus, A. P., Dou, J., *et al.* (2020). Machine learning methods for landslide susceptibility studies: A comparative overview of algorithm performance. *Earth-Science Reviews*, 207, 103225.

[14] Liu, X., Faes, L., Kale, A. U., *et al.* (2019). A comparison of deep learning performance against health-care professionals in detecting diseases from medical imaging: A systematic review and meta-analysis. *The Lancet Digital Health*, 1(6), e271–e297.

[15] Dodge, S., and Karam, L. (2017). A study and comparison of human and deep learning recognition performance under visual distortions. In *2017 26th International Conference on Computer Communication and Networks (ICCCN)* (pp. 1–7). Piscataway, NJ: IEEE.

[16] Pandian, A. P. (2021). Performance evaluation and comparison using deep learning techniques in sentiment analysis. *Journal of Soft Computing Paradigm (JSCP)*, 3(2), 123–134.

[17] Chan, H. P., Samala, R. K., Hadjiiski, L. M., and Zhou, C. (2020). Deep learning in medical image analysis. In Lee, G., and Fujita, H. (eds.), *Deep Learning in Medical Image Analysis: Challenges and Applications*, (pp. 3–21). Cham: Springer.

[18] Ahmed, Z., Mohamed, K., Zeeshan, S., and Dong, X. (2020). Artificial intelligence with multi-functional machine learning platform development for better healthcare and precision medicine. *Database*, 2020, baaa010.

[19] Kononenko, I. (2001). Machine learning for medical diagnosis: History, state of the art and perspective. *Artificial Intelligence in Medicine*, 23(1), 89–109.

[20] Shah, P., Kendall, F., Khozin, S., *et al.* (2019). Artificial intelligence and machine learning in clinical development: A translational perspective. *NPJ Digital Medicine*, 2(1), 69.

[21] Noorbakhsh-Sabet, N., Zand, R., Zhang, Y., and Abedi, V. (2019). Artificial intelligence transforms the future of health care. *The American Journal of Medicine*, 132(7), 795–801.

[22] Kusuma, S., and Jothi, K. R. (2022). ECG signals-based automated diagnosis of congestive heart failure using deep CNN and LSTM architecture. *Biocybernetics and Biomedical Engineering*, 42(1), 247–257.

[23] Öztürk, Ş., and Özkaya, U. (2021). Residual LSTM layered CNN for classification of gastrointestinal tract diseases. *Journal of Biomedical Informatics*, 113, 103638.

[24] Dua, M., Makhija, D., Manasa, P. Y. L., and Mishra, P. (2020). A CNN–RNN–LSTM based amalgamation for Alzheimer's disease detection. *Journal of Medical and Biological Engineering*, 40(5), 688–706.

[25] Begum, A., Dhilip Kumar, V., Asghar, J., Hemalatha, D., and Arulkumaran, G. (2022). A combined deep CNN: LSTM with a random

forest approach for breast cancer diagnosis. *Complexity*, 2022, 9299621, 1–9.

[26] Hassan, M., Awan, F. M., Naz, A., *et al.* (2022). Innovations in genomics and big data analytics for personalized medicine and health care: A review. *International Journal of Molecular Sciences*, 23(9), 4645.

[27] Ho, D., Quake, S. R., McCabe, E. R., *et al.* (2020). Enabling technologies for personalized and precision medicine. *Trends in Biotechnology*, 38(5), 497–518.

[28] Sunil Krishnan, G., Joshi, A., and Kaushik, V. (2021). Bioinformatics in personalized medicine. In Singh, V., and Kumar, A. (eds), *Advances in Bioinformatics* (pp. 303–315). Singapore: Springer.

[29] Ory, M. G., Adepoju, O. E., Ramos, K. S., Silva, P. S., and Vollmer Dahlke, D. (2023). Health equity innovation in precision medicine: Current challenges and future directions. *Frontiers in Public Health*, 11, 1119736.

[30] Nardini, C., Osmani, V., Cormio, P. G., *et al.* (2021). The evolution of personalized healthcare and the pivotal role of European regions in its implementation. *Personalized Medicine*, 18(3), 283–294.

[31] Aceto, G., Persico, V., and Pescapé, A. (2020). Industry 4.0 and health: Internet of things, big data, and cloud computing for Healthcare 4.0. *Journal of Industrial Information Integration*, 18, 100129.

[32] Yang, J., Li, Y., Liu, Q., *et al.* (2020). Brief introduction of medical database and data mining technology in big data era. *Journal of Evidence-Based Medicine*, 13(1), 57–69.

[33] Waring, J., Lindvall, C., and Umeton, R. (2020). Automated machine learning: Review of the state-of-the-art and opportunities for healthcare. *Artificial Intelligence in Medicine*, 104, 101822.

[34] Shilo, S., Rossman, H., and Segal, E. (2020). Axes of a revolution: Challenges and promises of big data in healthcare. *Nature Medicine*, 26(1), 29–38.

[35] Althnian, A., AlSaeed, D., Al-Baity, H., *et al.* (2021). Impact of dataset size on classification performance: An empirical evaluation in the medical domain. *Applied Sciences*, 11(2), 796.

[36] Pandya, S., Thakur, A., Saxena, S., *et al.* (2021). A study of the recent trends of immunology: Key challenges, domains, applications, datasets, and future directions. *Sensors*, 21(23), 7786.

[37] Paullada, A., Raji, I. D., Bender, E. M., Denton, E., and Hanna, A. (2021). Data and its (dis) contents: A survey of dataset development and use in machine learning research. *Patterns*, 2(11), 100336.

[38] Krawiec, P., Latoszek, W., Batalla, J. M., Bajenaru, L., and Dobre, C. (2023). Using blockchain to improve data management in the vINCI Platform. In *2023 24th International Conference on Control Systems and Computer Science (CSCS)* (pp. 613–618). Piscataway, NJ: IEEE.

[39] Hampel, H., Vergallo, A., Perry, G., Lista, S., and Alzheimer Precision Medicine Initiative. (2019). The Alzheimer precision medicine initiative. *Journal of Alzheimer's Disease*, 68(1), 1–24.

[40] Maletzky, A., Böck, C., Tschoellitsch, T., *et al.* (2022). Lifting hospital electronic health record data treasures: Challenges and opportunities. *JMIR Medical Informatics*, 10(10), e38557.

[41] Laukvik, L. B., Lyngstad, M., Rotegård, A. K., and Fossum, M. (2024). Utilizing nursing standards in electronic health records: A descriptive qualitative study. *International Journal of Medical Informatics*, 184, 105350.

[42] Kim, M. K., Rouphael, C., McMichael, J., Welch, N., and Dasarathy, S. (2024). Challenges in and opportunities for electronic health record-based data analysis and interpretation. *Gut and Liver*, 18(2), 201.

[43] Abbas, B. (2024). Frontiers unveiled: Radiology's pioneering role in critical care diagnostics through emerging technologies and innovations. *EasyChair Preprint*, https://easychair.org/publications/preprint/ftJV.

[44] Yenneti, S. (2024). Imaging research in clinical trials. *Journal of Healthcare and Life-Science Research*, 3(2), 97–103.

[45] Sayed, D., Chakravarthy, K., Amirdelfan, K., *et al.* (2020). A comprehensive practice guideline for magnetic resonance imaging compatibility in implanted neuromodulation devices. *Neuromodulation: Technology at the Neural Interface*, 23(7), 893–911.

[46] Shahin, M. H., Bhattacharya, S., Silva, D., *et al.* (2020). Open data revolution in clinical research: Opportunities and challenges. *Clinical and Translational Science*, 13(4), 665–674.

[47] Inan, O. T., Tenaerts, P., Prindiville, S. A., *et al.* (2020). Digitizing clinical trials. *NPJ Digital Medicine*, 3(1), 101.

[48] Harper, B., Smith, Z., Snowdon, J., DiCicco, R., Hekmat, R., and Getz, K. (2021). Characterizing clinical data management challenges and their impact. *Applied Clinical Trials*, 30(1/2), https://www.appliedclinicaltrialsonline.com/view/characterizing-clinical-data-management-challenges-and-their-impact.

[49] D'Ordine, K. (2023). HIPAA vs. Medical Research: Improving Patient Care Through Integration of Data Privacy and Data Access. Thesis, Bryant University. https://digitalcommons.bryant.edu/honors_data_science/9/.

[50] Shah, S. M., and Khan, R. A. (2020). Secondary use of electronic health record: Opportunities and challenges. *IEEE Access*, 8, 136947–136965.

[51] Sah, S. (2020). Machine learning: A review of learning types. *Preprints*, 2020070230. https://doi.org/10.20944/preprints202007.0230.v1.

[52] Antunes, R. S., André da Costa, C., Küderle, A., Yari, I. A., and Eskofier, B. (2022). Federated learning for healthcare: Systematic review and architecture proposal. *ACM Transactions on Intelligent Systems and Technology (TIST)*, 13(4), 1–23.

[53] Chalamala, S. R., Kummari, N. K., Singh, A. K., Saibewar, A., and Chalavadi, K. M. (2022). Federated learning to comply with data protection regulations. *CSI Transactions on ICT*, 10(1), 47–60.

[54] Thummisetti, B. S. P., and Atluri, H. (2024). Advancing healthcare informatics for empowering privacy and security through federated learning paradigms. *International Journal of Sustainable Development in Computing Science*, 1(1), 1–16.

[55] Sheller, M. J., Edwards, B., Reina, G. A., *et al.* (2020). Federated learning in medicine: Facilitating multi-institutional collaborations without sharing patient data. *Scientific Reports*, 10(1), 12598.

[56] Geng, J., Kanwal, N., Jaatun, M. G., and Rong, C. (2021). Did-eFed: Facilitating federated learning as a service with decentralized identities. In *Proceedings of the 25th International Conference on Evaluation and Assessment in Software Engineering* (pp. 329–335).

[57] Aouedi, O., Sacco, A., Piamrat, K., and Marchetto, G. (2022). Handling privacy-sensitive medical data with federated learning: Challenges and future directions. *IEEE Journal of Biomedical and Health Informatics*, 27(2), 790–803.

[58] Rauniyar, A., Hagos, D. H., Jha, D., *et al.* (2023). Federated learning for medical applications: A taxonomy, current trends, challenges, and future research directions. *IEEE Internet of Things Journal*, 11(5), 7374–7398.

[59] Seh, A. H., Zarour, M., Alenezi, M. *et al.* (2020). Healthcare data breaches: insights and implications. *Healthcare*, 8(2), 133.

[60] Jacobs, B., and Popma, J. (2019). Medical research, big data and the need for privacy by design. *Big Data & Society*, 6(1), 1–5.

[61] Ming, Y., and Zhang, T. (2018). Efficient privacy-preserving access control scheme in electronic health records system. *Sensors*, 18(10), 3520.

[62] Pussewalage, H. S. G., and Oleshchuk, V. A. (2016). Privacy preserving mechanisms for enforcing security and privacy requirements in E-health solutions. *International Journal of Information Management*, 36(6), 1161–1173.

[63] Yoon, J., Mizrahi, M., Ghalaty, N. F., *et al.* (2023). EHR-Safe: Generating high-fidelity and privacy-preserving synthetic electronic health records. *NPJ Digital Medicine*, 6(1), 141.

[64] Kanwal, T., Anjum, A., Malik, S. U., *et al.* (2021). A robust privacy preserving approach for electronic health records using multiple dataset with multiple sensitive attributes. *Computers & Security*, 105, 102224.

[65] Zhang, A., and Lin, X. (2018). Towards secure and privacy-preserving data sharing in e-health systems via consortium blockchain. *Journal of Medical Systems*, 42(8), 140.

[66] Sahi, A., Lai, D., and Li, Y. (2016). Security and privacy preserving approaches in the eHealth clouds with disaster recovery plan. *Computers in Biology and Medicine*, 78, 1–8.

[67] Kho, A. N., Cashy, J. P., Jackson, K. L., *et al.* (2015). Design and implementation of a privacy preserving electronic health record linkage tool in Chicago. *Journal of the American Medical Informatics Association*, 22(5), 1072–1080.

[68] Kumar, A., Kumar, R., and Sodhi, S. S. (2020). Intelligent privacy preservation electronic health record framework using soft computing. *Journal of Information and Optimization Sciences*, 41(7), 1615–1632.

[69] Liang, J., Qin, Z., Xiao, S., Zhang, J., Yin, H., and Li, K. (2020). Privacy-preserving range query over multi-source electronic health records in public clouds. *Journal of Parallel and Distributed Computing*, 135, 127–139.

[70] Lee, D., and Song, M. (2021). MEXchange: A privacy-preserving blockchain-based framework for health information exchange using ring signature and stealth address. *IEEE Access*, 9, 158122–158139.

[71] Lee, S., Kim, J., Kwon, Y., Kim, T., and Cho, S. (2022). Privacy preservation in patient information exchange systems based on blockchain: System design study. *Journal of Medical Internet Research*, 24(3), e29108.

[72] Dimitropoulos, L. L. (2007). Privacy and security solutions for interoperable health information exchange. *Report to the American Health Information Community. RTI International.*

[73] Lee, S., Kim, Y., and Cho, S. (2024). Searchable blockchain-based healthcare information exchange system to enhance privacy preserving and data usability. *Sensors*, 24(5), 1582.

[74] Mezher, M., Hellmuth, N., and Ibrahim, A. (2021). Secure health information exchange (S-HIE) protocol with reduced round-trip count. In *2021 IEEE 9th International Conference on Healthcare Informatics (ICHI)* (pp. 280–288). Piscataway, NJ: IEEE.

[75] Spanakis, E. G., Sfakianakis, S., Bonomi, S., Ciccotelli, C., Magalini, S., and Sakkalis, V. (2021). Emerging and established trends to support secure health information exchange. *Frontiers in Digital Health*, 3, 636082.

[76] Fadheel, W., Salih, R., and Lilien, L. (2018). PHeDHA: Protecting healthcare data in health information exchanges with active data bundles. In *2018 17th IEEE International Conference on Trust, Security and Privacy in Computing and Communications/12th IEEE International Conference on Big Data Science and Engineering (TrustCom/BigDataSE)* (pp. 1187–1195). Piscataway, NJ: IEEE.

[77] Esmaeilzadeh, P. (2018). The effects of public concern for information privacy on the adoption of health information exchanges (HIEs) by healthcare entities. *Health Communication*, 34(10), 1202–1211.

[78] Ibrahim, A. F. S. (2016). New secure solutions for privacy and access control in health information exchange. Theses and Dissertations, University of Kentucky, http://dx.doi.org/10.13023/ETD.2016.307.

[79] Esmaeilzadeh, P. (2022). Benefits and concerns associated with blockchain-based health information exchange (HIE): A qualitative study from physicians' perspectives. *BMC Medical Informatics and Decision Making*, 22(1), 80.

[80] Ghubaish, A., Salman, T., Zolanvari, M., Unal, D., Al-Ali, A., and Jain, R. (2020). Recent advances in the Internet-of-Medical-Things (IoMT) systems security. *IEEE Internet of Things Journal*, 8(11), 8707–8718.

[81] Rahman, A., Hossain, M. S., Alrajeh, N. A., and Alsolami, F. (2020). Adversarial examples—Security threats to COVID-19 deep learning systems in medical IoT devices. *IEEE Internet of Things Journal*, 8(12), 9603–9610.

[82] Hasan, M. K., Islam, S., Sulaiman, R., *et al.* (2021). Lightweight encryption technique to enhance medical image security on internet of medical things applications. *IEEE Access*, 9, 47731–47742.

[83] Elhoseny, M., Ramírez-González, G., Abu-Elnasr, O. M., Shawkat, S. A., Arunkumar, N., and Farouk, A. (2018). Secure medical data transmission model for IoT-based healthcare systems. *IEEE Access*, 6, 20596–20608.

[84] Martinez, J. B. (2018). Medical device security in the IoT age. In *2018 9th IEEE Annual Ubiquitous Computing, Electronics & Mobile Communication Conference (UEMCON)* (pp. 128–134). Piscataway, NJ: IEEE.

[85] Nomikos, K., Papadimitriou, A., Stergiopoulos, G., Koutras, D., Psarakis, M., and Kotzanikolaou, P. (2020). On a security-oriented design framework for medical IoT devices: The hardware security perspective. In *2020 23rd Euromicro Conference on Digital System Design (DSD)* (pp. 301–308). Piscataway, NJ: IEEE.

[86] Safkhani, M., Bendavid, Y., Rostampour, S., and Bagheri, N. (2019). On designing lightweight RFID security protocols for medical IoT. *Cryptology ePrint Archive*, Paper 2019/851, https://eprint.iacr.org/2019/851.

[87] Hasan, M. K., Ghazal, T. M., Saeed, R. A., *et al.* (2022). A review on security threats, vulnerabilities, and counter measures of 5G enabled Internet-of-Medical-Things. *IET communications*, 16(5), 421–432.

[88] Sengan, S., Khalaf, O. I., Priyadarsini, S., Sharma, D. K., Amarendra, K., and Hamad, A. A. (2022). Smart healthcare security device on medical IoT using raspberry pi. *International Journal of Reliable and Quality E-Healthcare (IJRQEH)*, 11(3), 1–11.

[89] Zakaria, H., Bakar, N. A. A., Hassan, N. H., and Yaacob, S. (2019). IoT security risk management model for secured practice in healthcare environment. *Procedia Computer Science*, 161, 1241–1248.

[90] Hale, T. M., and Kvedar, J. C. (2014). Privacy and security concerns in telehealth. *AMA Journal of Ethics*, 16(12), 981–985.

[91] Watzlaf, V. J., Zhou, L., DeAlmeida, D. R., and Hartman, L. M. (2017). A systematic review of research studies examining telehealth privacy and security practices used by healthcare providers. *International Journal of Telerehabilitation*, 9(2), 39.

[92] Ahmad, R. W., Salah, K., Jayaraman, R., Yaqoob, I., Ellahham, S., and Omar, M. (2021). The role of blockchain technology in telehealth and telemedicine. *International Journal of Medical Informatics*, 148, 104399.

[93] Zhou, L., Thieret, R., Watzlaf, V., DeAlmeida, D., and Parmanto, B. (2019). A telehealth privacy and security self-assessment questionnaire for telehealth providers: Development and validation. *International Journal of Telerehabilitation*, 11(1), 3.

[94] AlOsail, D., Amino, N., and Mohammad, N. (2021). Security issues and solutions in e-health and telemedicine. In *Computer Networks, Big Data and IoT: Proceedings of ICCBI 2020* (pp. 305–318). Singapore: Springer.

[95] Vega-Barbas, M., Seoane, F., and Pau, I. (2019). Characterization of user-centered security in telehealth services. *International Journal of Environmental Research and Public Health*, 16(5), 693.

[96] Pramesha Chandrasiri, G. A., Halgamuge, M. N., and Subhashi Jayasekara, C. (2019). A comparative study in the application of IoT in health care: Data security in telemedicine. In Mahmood, Z. (ed.), *Security, Privacy and Trust in the IoT Environment*, (pp. 181–202) Cham: Springer.

[97] Wenhua, Z., Hasan, M. K., Jailani, N. B., *et al.* (2024). A lightweight security model for ensuring patient privacy and confidentiality in telehealth applications. *Computers in Human Behavior*, 153, 108134.

[98] Sharma, H. K., Choudhury, T., Katal, A., and Um, J. S. (2022). Security and privacy issue in telemedicine: Issues, solutions, and standards. In *Telemedicine: The Computer Transformation of Healthcare* (pp. 185–196). Cham: Springer International Publishing.

[99] Chaet, D., Clearfield, R., Sabin, J. E., Skimming, K., and Council on Ethical and Judicial Affairs American Medical Association. (2017). Ethical practice in telehealth and telemedicine. *Journal of General Internal Medicine*, 32, 1136–1140.

[100] Arora, S., Yttri, J., and Nilsen, W. (2014). Privacy and security in mobile health (mHealth) research. *Alcohol Research: Current Reviews*, 36(1), 143.

[101] Zubaydi, F., Saleh, A., Aloul, F., and Sagahyroon, A. (2015). Security of mobile health (mHealth) systems. In *2015 IEEE 15th international Conference on Bioinformatics and Bioengineering (BIBE)* (pp. 1–5). Piscataway, NJ: IEEE.

[102] Dehling, T., Gao, F., Schneider, S., and Sunyaev, A. (2015). Exploring the far side of mobile health: Information security and privacy of mobile health apps on iOS and Android. *JMIR mHealth and uHealth*, 3(1), e3672.

[103] Tangari, G., Ikram, M., Sentana, I. W. B., Ijaz, K., Kaafar, M. A., and Berkovsky, S. (2021). Analyzing security issues of android mobile health and medical applications. *Journal of the American Medical Informatics Association*, 28(10), 2074–2084.

[104] Papageorgiou, A., Strigkos, M., Politou, E., Alepis, E., Solanas, A., and Patsakis, C. (2018). Security and privacy analysis of mobile health applications: The alarming state of practice. *IEEE Access*, 6, 9390–9403.

[105] Aliasgari, M., Black, M., and Yadav, N. (2018). Security vulnerabilities in mobile health applications. In *2018 IEEE Conference on Application, Information and Network Security (AINS)* (pp. 21–26).Piscataway, NJ: IEEE.

[106] Adhikari, R., Richards, D., and Scott, K. (2014). Security and privacy issues related to the use of mobile health apps. *Proceedings of the 25th Australasian Conference on Information Systems* (pp. 1–11). Auckland, New Zealand: ACIS.

[107] Plachkinova, M., Andrés, S., and Chatterjee, S. (2015). A taxonomy of mHealth apps—Security and privacy concerns. In *2015 48th Hawaii International Conference on System Sciences* (pp. 3187–3196). Piscataway, NJ: IEEE.

[108] Sun, J., Zhu, X., Zhang, C., and Fang, Y. (2011). Security and privacy for mobile healthcare (m-health) systems. In Das, S. K., Kant, K., and Zhang,

N. (eds.), *Handbook on Securing Cyber-Physical Infrastructure*. Waltham, MA: Morgan Kaufmann (pp. , 677–704).

[109] Hussain, M., Al-Haiqi, A., Zaidan, A. A., *et al.* (2018). A security framework for mHealth apps on Android platform. *Computers & Security*, 75, 191–217.

[110] Azeez, N. A., and Van der Vyver, C. (2019). Security and privacy issues in e-health cloud-based system: A comprehensive content analysis. *Egyptian Informatics Journal*, 20(2), 97–108.

[111] Khattak, H. A. K., Abbass, H., Naeem, A., Saleem, K., and Iqbal, W. (2015). Security concerns of cloud-based healthcare systems: A perspective of moving from single-cloud to a multi-cloud infrastructure. In *2015 17th International Conference on E-health Networking, Application & Services (HealthCom)* (pp. 61–67). Piscataway, NJ: IEEE.

[112] Raval, D., and Jangale, S. (2016). Cloud based information security and privacy in healthcare. *International Journal of Computer Applications*, 150(4), 11–15.

[113] JPC Rodrigues, J., de la Torre, I., Fernández, G., and López-Coronado, M. (2013). Analysis of the security and privacy requirements of cloud-based electronic health records systems. *Journal of Medical Internet Research*, 15(8), e186.

[114] Sharaf, S., and Shilbayeh, N. F. (2019). A secure G-cloud-based framework for government healthcare services. *IEEE Access*, 7, 37876–37882.

[115] Marwan, M., AlShahwan, F., Sifou, F., Kartit, A., and Ouahmane, H. (2019). Improving the security of cloud-based medical image storage. *Engineering Letters*, 27(1), 1–19.

[116] Aledhari, M., Marhoon, A., Hamad, A., and Saeed, F. (2017). A new cryptography algorithm to protect cloud-based healthcare services. In *2017 IEEE/ACM International Conference on Connected Health: Applications, Systems and Engineering Technologies (CHASE)* (pp. 37–43). Piscataway, NJ: IEEE.

[117] Benil, T., and Jasper, J. J. C. N. (2020). Cloud based security on outsourcing using blockchain in E-health systems. *Computer Networks*, 178, 107344.

[118] Masud, M., Gaba, G. S., Choudhary, K., Alroobaea, R., and Hossain, M. S. (2021). A robust and lightweight secure access scheme for cloud based E-healthcare services. *Peer-to-Peer Networking and Applications*, 14(5), 3043–3057.

[119] Khan, F. A., Ali, A., Abbas, H., and Haldar, N. A. H. (2014). A cloud-based healthcare framework for security and patients' data privacy using wireless body area networks. *Procedia Computer Science*, 34, 511–517.

[120] Nilsson, A., Smith, S., Ulm, G., Gustavsson, E., and Jirstrand, M. (2018). A performance evaluation of federated learning algorithms. In *Proceedings of the Second Workshop on Distributed Infrastructures for Deep Learning* (pp. 1–8).

[121] Lai, F., Dai, Y., Singapuram, S., *et al.* (2022). Fedscale: Benchmarking model and system performance of federated learning at scale. In

International Conference on Machine Learning (pp. 11814–11827). Baltimore, MD: PMLR.

[122] Gao, Y., Kim, M., Abuadbba, S., *et al.* (2020). End-to-end evaluation of federated learning and split learning for Internet of Things. In *International Symposium on Reliable Distributed Systems (SRDS)*, Shanghai, China (pp. 91–100).

[123] Wei, K., Li, J., Ding, M., *et al.* (2020). Federated learning with differential privacy: Algorithms and performance analysis. *IEEE Transactions on Information Forensics and Security*, 15, 3454–3469.

[124] Lee, G. H., and Shin, S. Y. (2020). Federated learning on clinical benchmark data: Performance assessment. *Journal of Medical Internet Research*, 22(10), e20891.

[125] Liu, Z., Chen, Y., Zhao, Y., *et al.* (2022). Contribution-aware federated learning for smart healthcare. In *Proceedings of the AAAI Conference on Artificial Intelligence* (vol. 36, no. 11, pp. 12396–12404).

[126] Wang, W., Li, X., Qiu, X., Zhang, X., Brusic, V., and Zhao, J. (2023). A privacy preserving framework for federated learning in smart healthcare systems. *Information Processing & Management*, 60(1), 103167.

[127] Sohan, M. F., and Basalamah, A. (2023). A systematic review on federated learning in medical image analysis. *IEEE Access*, 11, 28628–28644.

[128] Rajagopal, S. M., Supriya, M., and Buyya, R. (2023). FedSDM: Federated learning based smart decision making module for ECG data in IoT integrated edge-fog-cloud computing environments. *Internet of Things*, 22, 100784.

[129] Houssein, E. H., and Sayed, A. (2023). Boosted federated learning based on improved particle swarm optimization for healthcare IoT devices. *Computers in Biology and Medicine*, 163, 107195.

[130] Yaqoob, M. M., Nazir, M., Khan, M. A., Qureshi, S., and Al-Rasheed, A. (2023). Hybrid classifier-based federated learning in health service providers for cardiovascular disease prediction. *Applied Sciences*, 13(3), 1911.

[131] Kandati, D. R., and Gadekallu, T. R. (2023). Federated learning approach for early detection of chest lesion caused by COVID-19 infection using particle swarm optimization. *Electronics*, 12(3), 710.

[132] Anter, A. M., and Abualigah, L. (2023). Deep federated machine learning-based optimization methods for liver tumor diagnosis: A review. *Archives of Computational Methods in Engineering*, 30(5), 3359–3378.

[133] Cunha, M., Mendes, R., and Vilela, J. P. (2021). A survey of privacy-preserving mechanisms for heterogeneous data types. *Computer Science Review*, 41, 100403.

[134] Jurczyk, P., and Xiong, L. (2008). Towards privacy-preserving integration of distributed heterogeneous data. In *Proceedings of the 2nd PhD Workshop on Information and Knowledge Management* (pp. 65–72).

[135] Wang, X., He, J., Cheng, P., and Chen, J. (2018). Privacy preserving collaborative computing: Heterogeneous privacy guarantee and efficient incentive mechanism. *IEEE Transactions on Signal Processing*, 67(1), 221–233.

[136] Mohammed, N., Jiang, X., Chen, R., Fung, B. C., and Ohno-Machado, L. (2013). Privacy-preserving heterogeneous health data sharing. *Journal of the American Medical Informatics Association*, 20(3), 462–469.

[137] Zhang, L., Shen, L., Ding, L., Tao, D., and Duan, L. Y. (2022). Fine-tuning global model via data-free knowledge distillation for non-IID federated learning. In *Proceedings of the IEEE/CVF Conference on Computer Vision and Pattern Recognition* (pp. 10174–10183).

[138] Zheng, J., Chen, Y., and Lai, Q. (2024). PPSFL: Privacy-preserving split federated learning for heterogeneous data in edge-based Internet of Things. *Future Generation Computer Systems*, 156, 231–241.

[139] Trindade, S., Bittencourt, L. F., and da Fonseca, N. L. (2022). Resource management at the network edge for federated learning. *Digital Communications and Networks*, 10(3), 765–782.

[140] Shkurti, L., Selimi, M., and Besimi, A. (2023). FederatedMesh: Collaborative federated learning for medical data sharing in mesh networks. In *International Conference on Collaborative Computing: Networking, Applications and Worksharing* (pp. 154–169). Cham: Springer Nature.

[141] Brisimi, T. S., Chen, R., Mela, T., Olshevsky, A., Paschalidis, I. C., and Shi, W. (2018). Federated learning of predictive models from federated electronic health records. *International Journal of Medical Informatics*, 112, 59–67.

[142] Remedios, S. W., Butman, J. A., Landman, B. A., and Pham, D. L. (2020). Federated gradient averaging for multi-site training with momentum-based optimizers. In *Domain Adaptation and Representation Transfer, and Distributed and Collaborative Learning: Second MICCAI Workshop, DART 2020, and First MICCAI Workshop, DCL 2020, Held in Conjunction with MICCAI 2020, Lima, Peru, October 4–8, 2020, Proceedings 2* (pp. 170–180). Springer International Publishing.

[143] Jiang, M., Wang, Z., and Dou, Q. (2022). Harmofl: Harmonizing local and global drifts in federated learning on heterogeneous medical images. In *Proceedings of the AAAI Conference on Artificial Intelligence* (vol. 36, no. 1, pp. 1087–1095).

[144] Li, W., Milletarì, F., Xu, D., *et al.* (2019). Privacy-preserving federated brain tumour segmentation. In *Machine Learning in Medical Imaging: 10th International Workshop, MLMI 2019, Held in Conjunction with MICCAI 2019, Shenzhen, China, October 13, 2019, Proceedings 10* (pp. 133–141). Springer International Publishing.

[145] Otálora, S., Rafael-Patiño, J., Madrona, A., *et al.* (2022, September). Weighting schemes for federated learning in heterogeneous and imbalanced segmentation datasets. In *International MICCAI Brainlesion Workshop* (pp. 45–56). Cham: Springer Nature.

[146] Ma, J., Zhang, Q., Lou, J., Xiong, L., and Ho, J. C. (2021). Communication efficient federated generalized tensor factorization for collaborative health data analytics. In *Proceedings of the Web Conference 2021* (pp. 171–182).

[147] Darzidehkalani, E., Sijtsema, N. M., and van Ooijen, P. M. A. (2023). A comparative study of federated learning models for COVID-19 detection. arXiv:2303.16141, https://doi.org/10.48550/arXiv.2303.16141.

[148] Dayan, I., Roth, H. R., Zhong, A., *et al.* (2021). Federated learning for predicting clinical outcomes in patients with COVID-19. *Nature Medicine*, 27, 1735–1743.

[149] Singhal, V., Jain, S. S., Anand, D., *et al.* (2020). Artificial intelligence enabled road vehicle-train collision risk assessment framework for unmanned railway level crossings. *IEEE Access*, 8, 113790–113806.

[150] Jhanjhi, N. Z., Brohi, S. N., Malik, N. A., and Humayun, M. (2020). Proposing a hybrid RPL protocol for rank and wormhole attack mitigation using machine learning. In *2020 2nd International Conference on Computer and Information Sciences (ICCIS)* (pp. 1–6). Piscataway, NJ: IEEE.

[151] Gaur, L., Afaq, A., Solanki, A., *et al.* (2021). Capitalizing on big data and revolutionary 5G technology: Extracting and visualizing ratings and reviews of global chain hotels. *Computers and Electrical Engineering*, 95, 107374.

[152] Gaur, L., Singh, G., Solanki, A., *et al.* (2021). Disposition of youth in predicting sustainable development goals using the neuro-fuzzy and random forest algorithms. *Human-Centric Computing and Information Sciences*, 11, 24.

[153] Adeyemo, V. E., Abdullah, A., JhanJhi, N. Z., Supramaniam, M., and Balogun, A. O. (2019). Ensemble and deep-learning methods for two-class and multi-attack anomaly intrusion detection: An empirical study. *International Journal of Advanced Computer Science and Applications*, 10(9), 520–529.

Chapter 8

Splitfed learning for multimodal emotion detection

Gowtham M.[1] *and Chandru A.S.*[2]

Although machine learning has greatly benefited us, more data is still required for improved accuracy and performance. Sensitive information like a person's face and expression may be included in this data, and gathering such information poses a serious privacy concern. Understanding emotions is essential for positive social interactions. It also helps to better gauge how customers feel about a product and helps to understand how their behaviors are changing. For this reason, the market for emotion recognition is currently booming despite having a lot of gaps in data. We can decentralize the training process and safeguard sensitive data by using the split federated learning model that gets around federated learning's resource limitations and shortens split learning's computation time. Our goal is to use split federated learning to develop an emotion classifier. Split federated learning, which blends split and federated learning, is how our classifier is put into practice. Using the splitfed, our classifier produced accuracy values of 87%, 98%, 96%, 87%, and 99% for the FER2013plus, AffectNet, CKplus, ouluCASIA, and KDEF datasets, in that order. When it compared to the accuracy results of the centralized training strategy for the identical datasets, the results were comparatively good.

Keywords: Federated; Split; Learning; Decentralized; Emotions

8.1 Introduction

According to Baltrušaitis *et al.* [1] and Abdullah *et al.* [2], emotion plays a critical role in interpersonal interactions, knowledge perception, insight, and other facets of life. Since acquiring and experiencing emotions is necessary for inter-communication in social settings, understanding of emotions has become crucial for humans to function on a daily basis. With the growing need for automated emotion identification systems, emotion detection has become an important field of

[1]Department of Computer Science and Engineering, The National Institute of Engineering, Mysuru, India
[2]Artificial Intelligence & Machine Learning, Jyothy Institute of Technology, Bengaluru, India

study in the human-computer interactions [3]. EEG-based brain waves, speech, text, and facial signals have all been utilized to investigate emotion detection [4]. Emotion recognition technologies provide a means of promoting intuitive or acceptable user interactions, particularly in light of the expanding and varied applications of human-computer interaction. The market for emotion detection and recognition (EDR) was valued at USD 19.87 million in 2020 and is projected to grow at a compound annual growth rate (CAGR) of 18.01% to USD 52.86 million by 2026.

One can use emotional signals from several modalities to predict the emotional state of a subject. Conversely, the single modal model finds it difficult to determine the emotional state of the customer. Looking at an object or event in front of us alone does not reveal someone's emotional state. This is one of the main arguments in favor of treating emotion recognition as a multimodal issue. The process of developing emotion-specific traits is complicated by a number of reasons, including multi-dimensional data, non-linear interactions across multiple modalities, and evidence of individual variances in emotion presentation. It has been demonstrated that deep networks and machine learning (ML) can get over these restrictions by finding intricate non-linear feature correlations in a multimodal data.

Furthermore, it is crucial to select the most significant attributes from a large feature set, necessitating the use of dimensionality reduction techniques. Moreover, feature extraction and selection calculations are computationally intensive. For instance, the computing cost of feature selection may increase significantly when feature dimensionality rises. In nearly every area of computing, including natural language processing (NLP), machine translation, computer vision, audio recognition, and others, algorithms have improved significantly as a result of deep learning (DL) technologies like convolutional neural networks (CNN), recurrent neural networks (RNN), and autoencoders [5].

Recently, DL has been utilized to construct configurable emotion identification structures due to its ability to give high-level data abstraction. Deep neural networks are used in DL approaches to extract characteristic features from high-level data representation.

The majority of emotion acquisition systems use 2D, non-immersive visual aids, like pictures or films, to elicit different emotional states. Conversely, immersive virtual reality (VR) is becoming more and more popular in the field of emotion study because it makes it easier for researchers to simulate scenarios in controlled laboratory settings with a strong sense of presence and engagement. Additionally, by leveraging its implicit measures and integrating with ML techniques, it can be used to investigate solutions that could have a widespread impact on a variety of research topics, opening up new avenues for scientists. An abundance of information found in sensor data, such as gyroscope and accelerometer data, can be used to evaluate a user's social behaviors, including social communications, location, and physical activity.

The emotions that are being studied can generally be categorized as positive, negative, or neutral. While persistently negative emotions can be detrimental to people's health, well-being, and career opportunities, positive emotions can increase emotional and subjective well-being. Comparatively bland, neutral emotions lack both a pronounced pleasure and a defining pain.

Being able to distinguish between an experience that is merely satisfactory and one that is exceedingly happy essentially comes down to fine-grained emotion identification. The numerous emotion fusion techniques, emotion classification algorithms, and emotion acquisition techniques developed between 2013 and 2021 are all thoroughly examined in this study. We take into account the wealth of studies on subtle emotional states. We go into great detail on the current dataset collection for emotion recognition. Lastly, we emphasize research gaps and topics that need more study while summarizing the research findings from the papers that have been published. Since ML and DL techniques have shown to be effective in handling the laborious preprocessing and feature extraction stages involved in multimodal emotion recognition, they are the main focus of this work. They are more accurate in identifying emotions and perform better in classification than traditional techniques. New technical methods such as transfer learning, federated learning, and artificial intelligence can be used to assess human emotion states. These studies are still in their early phases, though further investigation is needed to solve a variety of difficulties such as privacy and error rates. These study's contributions and the remaining portions of the document are arranged as follows: In Section 8.2, the background and uses of emotion recognition are covered. The relevant research and survey articles on this topic are included in Section 8.3. The research article's review process is described in Section 8.4.

Section 8.6 covers the many instruments and technologies used to collect data for emotion detection; Section 8.7 offers a detailed overview of diverse datasets; and Section 8.5 examines emotion and its more subtle expressions that are relevant to emotion recognition. The several classifiers for emotion recognition are reviewed in Section 8.7. The several data fusion techniques for emotion identification are shown in Section 8.8. Section 8.9 discusses potential paths for future research.

8.1.1 Motivation

Personalization benefits us as users by making our lives more convenient and effective [6]. The issue is that more data is required for better research and customization in order to train the models for better performance and accuracy. However, gathering this data from users and centrally storing it to train the model poses a significant privacy risk, particularly for sensitive data such as financial, medical, and personal information like an individual's face and expression. Several well-known businesses have seized the chance to benefit from emotion detection [7]. Disney, for instance, developed an algorithm that recognizes complicated emotions and may even forecast future emotions to gauge how well its films are received by the public. Another such is Affectiva, which created sophisticated object and emotion identification for in-car safety systems that can identify whether the driver is angry, content, sleepy, or depressed. The COVID-19 epidemic has had a significant impact on our lives recently and has resulted in mask regulations. It is essential in assisting us in stopping the disease's spread. Before COVID, covering some of one's face was customary for cultural, religious, and even legal reasons, as well as because it was required in some occupations [8].

8.1.2 Problem synopsis

In this study, we will develop a split federated learning hybrid algorithm to safeguard user privacy while enhancing the effectiveness of the training process. The goal of the algorithm is to reduce split learning's calculation time and get over the resource limitations of federated learning. The results and conclusion of this study are finally outlined in Section 8.10.

8.2 Background section

The authors' summary of the current status of emotion recognition, a thorough analysis of the range of its most significant applications, and a multimodal emotion recognition status report are all included in this part.

8.2.1 Emotion recognition

The process of detecting and classifying human emotions is known as emotion detection, and it makes use of technological abilities such as DL, biosensors, speech, voice, face, and pattern recognition. It is arbitrary to interpret emotions at first from a self-assessment questionnaire, speech inflection, or facial expression. Furthermore, it reduces recognition accuracy because of biases in expressions, gestures, or words. People could engage in intentional efforts to conceal or mimic their actual emotions. Such identification of emotions based on human judgment may be subject to human mistake. Thus, physiological signal measurements utilized in emotion recognition approaches are becoming more and more common, electromyogram, including galvanic skin reaction and heart response.

Modalities reflect a variety of information sources that might provide different types of information and perspectives. There are two types of emotion recognition algorithms: unimodal and multimodal. Human emotions can be recognized by a single modality such as a face, text, EEG, speech, or image, utilizing unimodal emotion recognition. The primary drawback of the unimodal is that, depending on the situation, the selected modality may not work well. Every modality has advantages and disadvantages. When predicting valence dimension in the low-light conditions, audio is better than video, while text is sometimes better than audio. The system requires all of the available data in order to obtain a global picture. In multimodal emotion identification, distinct modalities can predict various emotional states. In order to address the weaknesses of each modality and create a more resilient system, the goal of multimodal emotion recognition is to employ the combined information.

8.2.2 Multimodal recognition of emotions

Multimodal recognition of emotions mimics human prediction ability by integrating behavioral and biological characteristics. Multimodal emotion recognition necessitates the simultaneous processing of numerous data sources, in contrast to unimodal emotion recognition. Even among multimodal methodologies, there is a great deal of variation in prediction accuracy, which calls for the creation of reliable methods [9].

According to Song and Kim [10], hidden emotions are those in which the biosignal is active but the facial expression is neutral. Additionally, by evaluating feature maps of video and EEG inputs using CNN, their method effectively recognized concealed emotions. First, when the participant deliberately concealed a feeling, the exterior and internal emotional characteristics were assessed using multimodal signals. Then, using CNNs, which have powerful cognitive abilities, a method for detecting hidden emotions is devised based on the difference between internal and external emotions. This research has advanced the technique for deeply comprehending inner feelings. The many phases involved in multimodal recognition of emotions identification are described in Figure 8.1.

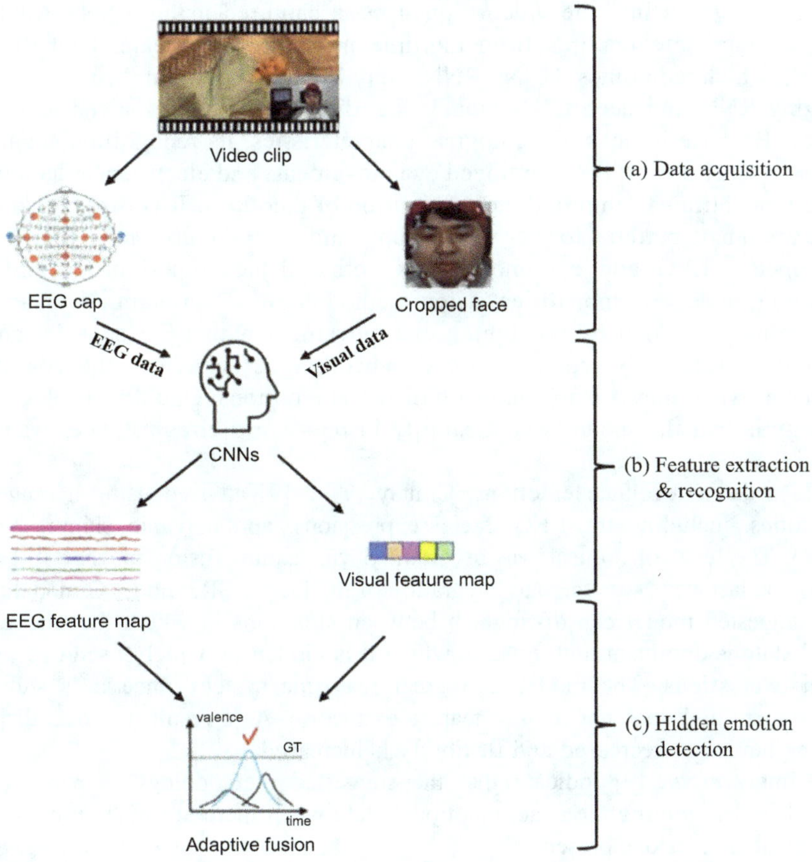

Figure 8.1 Multimodal recognition of emotions [11]

Povolny *et al.* [12] proposed a multimodal emotion identification system that uses both audio and visual information. The given audio clip was enhanced using bottle-neck (BN) features created by a constrained hidden layer of a neural network trained toward phonetic goals.

They used a CNN trained to recognize facial landmarks in order to enhance the baseline video data. These activations integrate geometrical information with appearance information. Furthermore, text-based features generated by an artificial speech recognition system are experimented with. The main benefit is that BN features have been shown to be highly successful in a number of audio processing applications, along with generic feature extraction techniques based on deep neural networks. The recognition accuracy will be below average since the technique lacks physiological properties.

Poria *et al.* [13] used voice, text, and facial gestures to perform multimodal sentiment analysis and emotion recognition. To extract textural and visual information, they employed a special temporal deep neural network. They merged the text, audio, and video multi-modal heterogeneous content they had retrieved using multiple kernel learning (MKL). The spatiotemporal information incorporated into the video sequences is captured using RNNs. MKL is used to aggregate features from multiple modalities to create the CRMKL model, which combines CNN, RNN, and MKL. The main benefit is that merging RNN and deep CNN models significantly increased speed and efficiency. Because it lacks physiological characteristics, its recognition accuracy will be poor. Lu *et al.* [14] employed eye movements and electroencephalogram (EEG) modalities to improve the recognition of emotions. It is discovered that there are innate patterns for negative, happy, and neutral emotions in the 16 eye movements. EEG and eye movements both enhance emotion recognition, according to several modalities fusion methodologies. The main advantage of this technique is that by combining eye movement with EEG data, it can be utilized to accurately represent users' subconscious behaviors and cognitive processes when they are aroused with different emotions. One drawback of this strategy is that the model's applicability in real-world circumstances was not tested.

In order to evaluate emotions, Cimtay *et al.* [9] advised using a range of modalities, including the EEG, facial expressions, and galvanic skin response (GSR). The level of arousal was ascertained via feature fusion on the GSR and EEG. The last stage was the late integration of the face, GSR, and EEG modalities. The suggested model can distinguish between situations in which the true emotional state is dominant and those in which it is hidden by widely used deceptive facial expressions. The model is a great real-time system since it is subject-independent and does not require feature extraction. As a result, the overall processing burden is decreased and flexibility is increased.

Cimtay *et al.* [9] indicate that the suggested methodology is particularly helpful in determining the exact emotional state when there are inherently deceptive facial expressions present.

8.2.3 Emotion recognition applications

In order to design and automate personalized interfaces or sub-component technologies for bigger systems, emotion detection techniques can be applied to video games, business intelligence and merchants, smart devices, car automation, e-learning, social robots, and virtual assistants. Below is a discussion of the primary application areas.

- **Learning online:** During the COVID-19 pandemic, online learning systems enable uninterrupted instruction, and real-time emotion recognition technology helps identify students' emotional states. With this understanding, lesson plans may be created to take into account the various learning preferences of the students [15].
- **Retail markets**: With the aid of Google Glass, Emotient, a company that specializes in emotion recognition, has been hoping to make a big impression in the retail sector. With the aim of leveraging this data to enhance the in-store experience, retailers will be able to observe what their customers are thinking and feeling. Its objective is to determine if emotion detection technologies can complement or perhaps replace standard satisfaction measurements by comparing data from these tools with data from standard assessments.
- **Uses in medicine:** Multimodal patient data is gathered in surveillance settings using automated IoMT equipment. The most recent facts and statistics are confirmed from relevant portals and papers for further research on various techniques for analyzing emotions. According to Zhang *et al.* [16], the proposed innovative IoMT quality service offers emotion-aware decision-making and real-time monitoring during the COVID-19 pandemic, significantly speeding up the process and delivering ongoing emotion-aware medical care.
- **Gaming:** Video game emotion detection research has the potential to increase player retention and enhance the overall gaming experience. To keep players' attention without interfering with their gaming experience, automatic emotion recognition is necessary [17]. Notwithstanding problems with latency and display quality, VR consoles give its users the impression that they are physically within the game. Since it enables game creators to swiftly adapt their games to a subscriber's feelings, emotion detection is a desirable feature to VR.
- **Robots with social skills:** Sentiment analysis and human-machine interaction are two other large topics that gained interest due to the ability to discern emotions. Due to their ability to perceive and understand human emotions, social robots are more beneficial in interpersonal interactions. They should be able to identify human emotions and modify their behavior in order to respond appropriately to them, claim Spezialetti *et al.* [18].

8.3 Related work

The authors examine current surveys in the field of emotion recognition in this section, outlining the approaches taken, the limits, and the taxonomy that was adhered to. Numerous studies have looked into techniques to get high recognition accuracy for emotions. As far as emotion recognition research survey studies go, Imani and Montazer [19], Sharma and Dhall [20], and Jam *et al.* [21] are the most pertinent studies currently available. Sharma and Dhall [20] outlined the advantages of emotion recognition in text, voice, picture, and EEG. Numerous datasets are examined, and both fusion techniques are well discussed. But the poll solely takes into account unimodal approaches.

Imani and Montazer [19] assessed a variety of emotion identification technologies utilized in e-learning platforms and enumerated their advantages and disadvantages. However, the mention of methods for emotion acquisition is omitted. The primary goal of Jam *et al.*'s [21] DL research was to discover the social and emotional aspects of human-robot interactions. The report requires more investigation on the relationship between emojis and in-depth emotional descriptions.

A thorough taxonomy of text emotion identification techniques was presented by Alswaidan and Menai [3], along with a thorough explanation of the key components and various methods that are accessible. Different methodologies identified in the literature are presented, along with their key characteristics, benefits, drawbacks, and comparisons, and a summary of many explicit and implicit emotion recognition strategies in the text is provided. However, tools and datasets are not covered in the survey. Several multimodal emotion identification techniques were presented by Abdullah *et al.* [2]. These techniques integrate various modalities, including text and image, as well as facial expressions and physiological reactions from the body. The results of this survey showed that integrating different modalities to identify emotions increases the accuracy of emotion recognition. A five-tiered taxonomy was presented by Baltrušaitis *et al.* [1], taking into account the difficulties that arise in all multimodal emotion identification techniques. Fusion, alignment, representation, co-learning, and translation are these five problems. In their study, co-learning was mentioned as a topic worth investigating. That being said, the majority of this overview paper focused on multimodal research conducted in the previous ten years.

The primary focus of Lim *et al.* [22] was on the classifiers and eye tracking techniques applied to this data. This study emphasizes how important it is to combine eye tracking technology with physiological signals. Nevertheless, there was an improper discussion or comparison of the survey's results between intra- and inter-subject classification performance rates. CNN-based speech emotion recognition was used to classify the three primary components of their work: voice data, extraction, feature, and DFC taxonomy (classification).

This makes it more likely that the speech's emotion will be accurately identified. The survey has a very small number of modalities because it exclusively focuses on speech emotion, making it exhaustive in its domain. The main problem with existing surveys that literature study has found is that the two studies that were previously discussed did not stress the significance of techniques for evaluating fine-grained emotions, emotion acquisition tools, or data fusion methodologies. Our study intends to offer a thorough and rigorous evaluation of the literature on various tools for acquiring emotions, datasets, fusion approaches, multimodal emotion identification classification models, and fine-grained emotion analysis. Finally, we summarize the results of the previous study and point out any knowledge gaps that might still remain in addition to potential future research areas.

8.4 Review method

This section provides a comprehensive review of the most recent research findings and outlines the current study areas in the subject of emotion recognition, summarizing the academic work that has been done in this area.

8.4.1 Methodology for searching

The following search terms were employed in accordance with the recommendations of this systematic literature review: ML, DL, emotion recognition, and multimodal emotion recognition. First, we automatically searched for conference and journal publications using Google Scholar. After that, we manually searched digital databases from Springer, IEEE, ACM, and Elsevier. The subsequent search terms were employed:

- "Recognizing Emotions."
- "Multimodal Identification of Emotions."
- "Emotion Identification" AND ("Intense Learning" OR "Computer Learning").
- "Recognition of Multimodal Emotions AND ("Deep Learning" OR "Machine Learning").

8.4.2 Research inquiries

Investigated in this study were the following research questions (RQs):

RQ 1: How can the various emotions be examined and classified?
RQ 2: Which methods and techniques are available for gathering information for emotion recognition?
RQ 3: What are the primary characteristics of the many emotion recognition datasets that are available?
RQ 4: What are the various techniques for classifying emotions that can be used to improve recognition accuracy?
RQ 5: Which data fusion methods ensure a higher level of accuracy in identifying emotions?
RQ 6: What are the obstacles and potential directions for further study that can improve the application of emotion recognition methods?

Our first research question (RQ 1) is addressed in Section 8.5, which also provides specific details about emotion and its different manifestations. In

Section 8.6, the second research question (RQ 2) is addressed. The several tools and technologies used to gather information to help in emotion recognition are examined in detail, along with their characteristics. In addition, Section 8.6 addresses the third research question (RQ 3) by providing and analyzing several datasets for emotion recognition. Our fourth research question (RQ 4) is outlined in Section 8.7, along with several examples of emotion classifiers. Our fifth research question (RQ 5), which outlines the numerous data fusion techniques used for emotion recognition, is summarized in Section 8.8. In order to respond to the final research question, Section 8.9 discusses the prospects and problems that the field of emotion recognition may face in the future (RQ 6).

8.5 Emotions

This section includes our first research question (RQ 1) and a description of the characteristics of emotion in all of its facets. Emotion is a conscious, subjective feeling that people experience in reaction to particular stimuli. It is a fundamental element of natural social interaction. An individual will unavoidably experience a wide variety of emotions during the day, ranging in sort and intensity. Emotions can also be understood as a person's response to environmental changes that occur on both a psychological and physical level. These emotions can range widely, from happiness to sadness to fury to excitement, and so on. Due to the complexity of human emotions, different situations have distinct effects on the emotions of different people. As multimedia and human-computer interaction technologies advance, the capacity to automatically identify people's emotional states becomes increasingly important. For example, a computer game's complexity can be changed by figuring out how the user feels.

The system will display a simple and entertaining game to a player who is feeling down in order to assist him decompress and kill time. The game difficulty rises when players are feeling good, which makes it harder for them to play. Emotional modification is also possible with music and film recommendations. Good multimedia material might lessen the effects of bad emotions whether watching movies or listening to music. Emotions are usually connected in some way; for example, sadness often involves a tiny amount of fury. Therefore, it's worthwhile to investigate how to combine emotional connections to create a more powerful emotion recognition model.

There are two categories of emotions: neutral and non-neutral. In general, neutral emotions are feelings that are indifferent to any given scenario. Non-neutral emotions come in two varieties: positive and negative. Negative characteristics, such as fear, failure, and despair, have been connected to negative emotions. Regularly going through these negative emotions can only be detrimental to one's health. It was also found to be directly related to an individual's attention span. An individual's heart system may potentially malfunction as a result of these negative emotions. Contrarily, positive emotions are experienced quite differently. Two positive feelings that are thought to be indicative of the best possible well-being are joy and happiness (Figure 8.2).

Positive and negative emotions have a clear association, as indicated by Shirke *et al.* [23]. A multimodal emotion recognition system that plays a film to elicit happy feelings and detects negative emotions is created using Brain IoT. When compared to other multimodal emotion identification techniques, the method's accuracy is lacking. Hipson *et al.* [24] made a clear distinction between being alone and lonely isolation. When one is alone, solitude is viewed as a nice state as opposed to the unpleasant experience of loneliness. NLP is used to examine tweets in order to ascertain the feelings they express. The results don't affect the overall tone of a tweet; they simply deal with the sentiment of words that are used in the same sentence as isolation phrases. Using neuroscan technology, Subasi *et al.* [25] classified the emotions as good, negative, and neutral. With the proposed method, errors in signal processing can be readily managed. Conversely, bigger and more diverse datasets might be used to assess this approach. Bosch *et al.*'s [26] study from 2022 concentrated on the negative feeling of dissatisfaction.

When comparing multilevel models, a system with individual expressions estimates expression frequency far better than a model with just the expression. Using facial image analysis, Suja and Tripathi [27] examined a variety of emotions, such as happiness, surprise, rage, disdain, and neutrality. This technique is used to detect emotions in real time using an image of a face. Their strategy is less accurate than other approaches [28]. Lu *et al.* [14] distinguished between positive, negative,

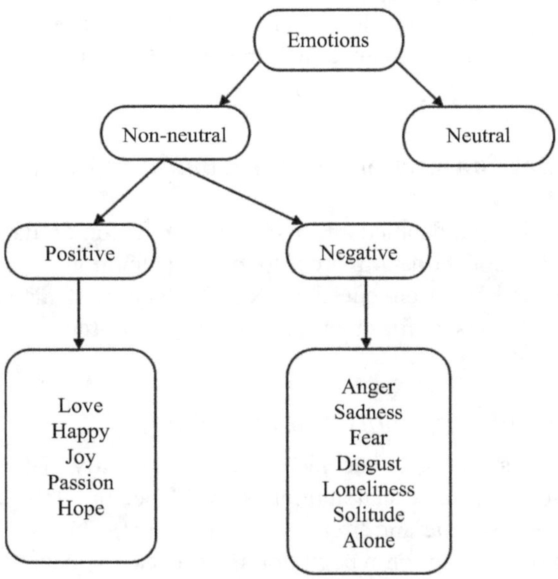

Figure 8.2 Taxonomy of emotions

and neutral emotions using eye gaze and EEG. The ability to distinguish between distinct emotional states can be enhanced by combining EEG with eye movements. Finding an emotion's "ground truth" might be difficult since emotions have fuzzy boundaries. List the various fine-grained emotions that are identified by various emotion recognition techniques. Recently, in an effort to capture a greater variety of emotional experiences, researchers have deliberately moved away from the six basic emotions. According to Marsella *et al.* [29], computational emotions can represent up to 25 distinct types of emotion and demonstrate high-dimensional emotional activity. Specific categories of emotion, rather than valence and arousal, are what drive the representation of emotion in experience, expression, and brain processing. Most affective reactions are found to be consistently merged instead of discrete. In an attempt to understand the phenomenon of emotions, software programs known as computational models of emotion assess emotional stimuli, elicit emotional reactions, and produce emotional actions using computational principles. Computational models of emotion are often based on psychological conceptions of emotion. At its most fundamental, the concept aims to define what an emotion is, for example, what defines an angry emotional state. Understanding human emotion is increasingly reliant on the research of computer models of emotion, which facilitate human-machine interaction. The importance of emotion in human behavior has been thoroughly demonstrated by modern research on human emotion; however, this reality has occasionally been overlooked in the fields of cognitive science and artificial intelligence. Because of this, researchers are already employing computational emotion models as instruments to examine human emotion and apply it to real-world situations. Virtual individuals are self-contained, embodied entities that are capable of verbal and nonverbal communication with actual people in person. Computational models of emotion are used in their creation.

8.6 Data collection and emotion recognition datasets

In addition to providing a detailed analysis and information on the many tools and technologies used to collect data in order to improve emotion recognition, this part addresses the second research question (RQ 2). This part also introduces and investigates several datasets for emotion recognition in order to address the third research question (RQ 3).

8.6.1 Data acquisition tools and technology

Emotion detection technologies become more commercially viable when used for specific applications such as visual-auditory surveillance or intelligent home robots that facilitate compassionate and lifelike human contact. This is feasible since the information needed for emotion recognition may be easily gathered with inexpensive equipment. Based on two classification categories, we created a taxonomy of emotion detection technologies and data gathering techniques (see Figure 8.3).

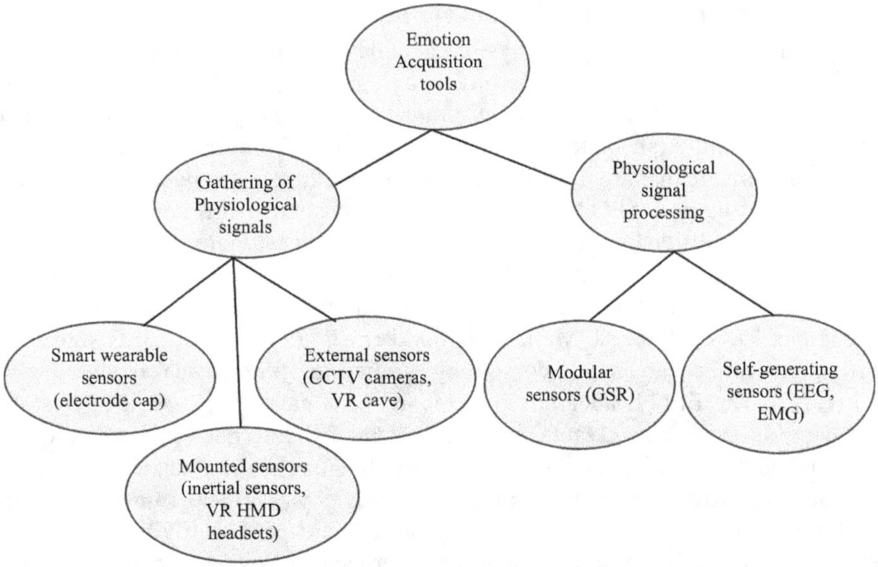

Figure 8.3 Taxonomy of emotion detection technologies and data gathering techniques

Sensors are in the first category since they are required for the development of emotions. Sensors are categorized as self-generating or modifying according to how they infer emotions from physiological data. Self-generating sensors record physiological information as a result of emotional emotions that independently set off particular bodily functions. Their output bandwidth corresponds with the quantity being measured, and they deliver output voltages or currents that are proportionate to the measured quantity. For instance, EEG sensors are self-generating gadgets that can recognize emotions without the need for an outside stimulus. To process emotions, modulating sensors need an external source. The GSR modulating sensor operates automatically to track skin conductance. But sympathetic activity, which governs a host of other human activities as well as mental and emotional states, is in charge of it.

The tools used to collect physiological data are the second primary classification group in our taxonomy. Wearable smart gadgets are attached to the body in order to quickly and accurately detect emotions. Electrode caps with dry or wet electrodes can be worn to record EEG data and identify emotions. Tools that are mounted to the body do so with the assistance of an external component. A multi-sensor mask that records a user's face movements, muscular signals, and physiological reactions was developed in 2021 by Gjoreski *et al.* [30]. Because they enable fully immersive systems that shut out extraneous inputs, head-mounted displays (HMDs) for VR are becoming more and more popular. External tools are gadgets that are used to elicit emotions through non-immersive stimuli; they are not physically affixed to the human body. This includes virtual caverns, CCTV, and cameras.

8.6.1.1 Sensors for physiological signal processing

Self-generating sensors capture physiological data resulting from emotional states that independently trigger specific physiological processes. They provide output voltages or currents that are commensurate to the quantity being measured, and their output bandwidth matches that of the quantity being measured. The self-generating sensors that are most frequently utilized are electromyography (EMG), electrodermal activity (EDA), and EEG. EEG sensors record noninvasive brain impulses, and the signals are analyzed to identify the best feature combination for identifying emotions. Nakisa *et al.* [31] classified an emotional response to positive and negative states using the characteristics of the EEG signal. Using a 14-channel EEG apparatus, Majid Mehmood *et al.* [32] captured EEG signals as participants looked at images containing elicitation stimuli such as happy, sad, and fearful. The EEG headcap used to record EEG data. Tang *et al.* [33] list the following as the main advantages of EEG: (1) EEG is not affected by either stimulus, and emotions can be hidden by language and facial expressions. (2) Compared to spoken and facial expressions, physiological signals are more reliable sources of information. (3) EEG signals are particularly useful since it gets easier and easier to gather users' physiological data as wearable computing technologies advance. (4) In multimodal retrieval, an individual's emotions can be represented via EEG signals. EMG, a technique for tracking muscle activity, has been linked to a range of emotional states. In order to measure the electrical activity of the face muscles during contraction, Hassan *et al.* [34] used zygomaticus electromyography (zEMG), which captures the major aspects of induced emotion. The electrical conductivity of the skin on the face, namely in the palmar areas, can be used to quantify psychological phenomena such as human cognitive and affective processes. Phasic and tonic components make up EDA. Shukla *et al.* [35] claim that skin conductance level (SCL) is an EDA tonic component.

Modulating sensors, in contrast to self-generating ones, need an outside source. Skin conductance is measured using the unconsciously modifying GSR sensor. It is autonomously governed by sympathetic activity, which influences mental and emotional states as well as behavioral elements of people. The GSR is a very accurate measure of emotional arousal. Usually, the palms or fingers are used. It controls the quantity of sweat secreted by skin pores. Human consciousness does not control GSR. Sweating is a way for the sympathetic nervous system to indirectly control human emotion. The photoplethysmogram (PPG), another modulating sensor, is a noninvasive, low-cost optical sensor that measures blood flow in the skin using infrared light. It also utilizes heart rate data to help identify emotions.

8.6.1.2 Instruments for acquiring emotions to collect physiological signals

Smart Wearables: By putting smart wearables on the body, emotions may be quickly and accurately sensed. Electrode caps with dry or wet electrodes can be worn to record EEG data and identify emotions. Using an Empatica wrist band, Ragot *et al.* [36] tracked cardiac activity and EDA signals. The scientists found that wearable sensors were exactly as capable as lab sensors at identifying moods. The principal limitation of this approach is that it only extracts a small number of variables for emotion identification.

EDA, PPG, and zEMG sensors were utilized by Hassan *et al.* [34] to measure physiological modalities. It facilitates the extraction of true emotions and gets over the limitations of functional neuroimaging-based emotion detection systems. The proposed paradigm struggled to discern between the four emotional states of happiness, sadness, relaxation, and disgust and neutral feelings.

Quiroz *et al.* [37] used a smart wearable watch equipped with a triaxial accelerometer and gyroscope to capture physiological information such as heart rate and pressure. This method's utilization of physiological features improved detection accuracy. One of the mentioned worries is the integrity of sensor data. While this technique can be used for real-time applications, it is not well suited for complex neural networks. Tools for generating feelings: Mounted instruments have an external component that holds them to the body. For emotion recognition, HMDs are the most widely used mounted devices. These gadgets enable the instant creation of a virtual world that our senses can experience. It creates the illusion that one is in the real world by using virtual experiences. Because it allows for the exclusion and adjustment of factors that would be impractical to carry out in real space, VR saves time and money. VR also makes it feasible to recreate and assess spatial settings under closely monitored experimental situations. Vergara *et al.* [38] divided VR applications into two primary categories according to the visualization and interaction tools employed.

- Non-immersive (the most popular window): the user views the world via the "window" of a computer's flat screen.
- Immersive refers to a method of virtual environment immersion wherein a wearer of glasses experiences the surroundings through two small screens placed in front of their eyes.
- Marín-Morales *et al.* [39] identified two types of VR immersive applications based on visualization technology.
- The term "immersive" describes a technique for creating a virtual environment where a person wearing glasses sees their surroundings through two tiny screens that are in front of their eyes.
- Stereoscopic projectors are used by the virtual CAVE (cave automated virtual environment) to project the virtual world onto the walls, ceiling, and floor of a space. In this instance, in order to see the virtual environment in three dimensions, the user will require passive stereo glasses. Numerous VR systems are presently available, varying in terms of cost and degree of user immersion.

8.6.2 Multimodal emotion recognition datasets

Large datasets are required to develop benchmark algorithms for automatic emotion recognition and to learn more about human emotional experience and expression. A number of emotion recognition datasets are collected from one or more modalities and made available to the general audience. This section displays the top nine emotion recognition datasets. Database EmoFBVP: Actor responses to affective emotion labels were captured by EmoFBVPD using their body language, facial expressions, vocalizations, and physical cues. Along with the multimodal recordings, this also includes information on skeletal tracking and face feature tracking. After every emotionally charged scene, the actors rate all of the data recorded on an assessment sheet.

DEAP: The public can access a multimodal dataset for assessing human emotional states. The Database for Emotion Analysis Using Physiological Signals is the name of the database. While watching clips of 41-minute music videos, 32 individuals had their EEG and peripheral physiological measurements recorded. The participants rated each movie according to dominance, valence, arousal, like/dislike, and familiarity. According to Koelstra *et al.* [40], 22 of the 32 participants had their frontal faces photographed.

SEED: The EEG signals recorded during the emotional induction process of subjects watching movie clips are included in the SEED (SJTU Emotion EEG) dataset. The movie scenes were carefully selected to elicit a range of feelings such as happiness, negativity, and neutrality. Fifteen Chinese film clips representing a range of emotional states were chosen as study stimuli from the available materials. The following criteria are used to choose the video clips: the movies should only evoke the one desired target feeling, the videos should be easy to watch and comprehend, and the experiment shouldn't be too long as this could tire out the participants. After watching each video clip, participants were invited to fill out a form discussing their emotional experiences.

FER2013: FER2013 was developed by using the Google image search API to look for pictures of faces that corresponded to a list of 184 emotion-related terms such as "blissful," "enraged," and so forth. These keywords yielded almost 600 strings of facial photo search searches when combined with terms pertaining to age, gender, and race. Each query's initial 1,000 photos were stored for the following processing cycle. The dataset contains 35,887 face photographs in total, the vast majority of which were taken outside. Happiness is represented by 8,989 photos, grief by 6,077 images, surprise by 4,002 photographs, rage by 4,953 images, scorn by 547 images, emotional neutrality by 6,198 images, and fear by 5,121 images. The age, posture, and occlusion conditions of faces vary widely. Create a challenging dataset. Furthermore, human recognition accuracy of about 65% is predicted [41].

K-EmoCon: K-EmoCon is a multimodal dataset that is perfect for researching emotions in social situations since it offers comprehensive descriptions of consistent emotion during realistic talks. The dataset includes multimodal metrics that were obtained during 16 sessions of ten-minute debates on a social topic using off-the-shelf approaches, together with video recordings, EEG, and peripheral physiological signals. The inclusion of emotion ratings from the viewpoints of oneself, one's argument partner, and external observers makes it apart from previous datasets. While watching a discussion clip, raters noted emotional displays for arousal-valence and 18 other group emotions every five seconds. Multiperspective emotional assessment of interpersonal connections is possible with the K-EmoCon dataset.

PMEmo: PMEmo is a multimodal research tool for emotion recognition and retrieval from public music. PMEmo consists of shared MER traits, manually chosen chorus sections, and song-level components. A large collection of music, standard valence-arousal annotations, real-time physiological inputs, and annotator recruitment components are all included in PMEmo. In total, 457 participants annotated 794 music samples in this dataset.

MAHNOB-HCI: In order to identify emotions and investigate implicit tagging, responses to emotional states are recorded in a multimodal database known as MAHNOB-HCI. A multimodal setup was used to synchronize eye gaze data, verbal cues, facial films, and physiological indications. In the two trials, there were 27 male and female subjects of various ethnic backgrounds. In the first test, participants saw 20 highly emotional movies, and then self-reported their feelings using criteria like valence, arousal, predictability, and dominance. In the second trial, short films and pictures were displayed first without tags and subsequently with the right or wrong ones. Participants were asked to score how much they agreed or disagreed with the displayed tags.

VREED: 360-degree immersive video-based virtual environments (360-VEs), delivered through a VR headset, were employed by the multimodal affective dataset VREED (VR Eyes: emotions dataset) to evoke feelings in users. The VREED contains data from the self-reported survey, eye tracking, ECG, and GSR. Data from 408 trials were submitted by 34 healthy volunteers in total. To choose the 360-VEs, focus groups and a pilot study with 12 more volunteers were employed. The entire dataset includes the technique, instructions, and samples from the questionnaire. According to the literature on emotional computing, a preliminary ML study was conducted and the results demonstrate state-of-the-art performance with non-immersive modalities. VREED, one of the earliest multimodal VR datasets, uses behavioral and physiological clues to identify emotions.

MEmoR: A new dataset for multimodal emotion reasoning based on videos is called MEmoR. MEmoR is a tool that provides fine-grained emotional annotations to both speakers and non-speakers. The well-known television program The Big Bang Theory provides 8,536 data samples and 5,502 video segments that MEmoR is constructed using. Fourteen emotions from Plutchik's wheel are used to annotate MEmoR at the individual level [9]. In these movies, speakers could not have as much visual detail as non-speakers, who are usually visually indistinguishable and devoid of any audio-textual cues. MEmoR is a more intricate and realistic testbed. In order to account for varying types of reasoning, the suggested MEmoR dataset incorporates both external and short-term data sources.

A subject who is asked to convey a particular emotion is said to be experiencing posed emotion. Spontaneous emotion is the subject's natural response to the elicitation.

8.7 Emotion recognition classifiers

Our fourth research question (RQ 4) is covered in this part, along with multiple learning strategies (ML/DL) for multimodal emotion classification. Classifiers, also referred to as pattern recognition techniques, categorize features that are applied to various sets of emotional inputs. There are two types of emotion classifiers: DL and basic ML. Figure 8.4, which depicts three areas in emotion recognition—classifier types, learning techniques employed, and applications—presents the taxonomy of emotion recognition classifiers.

8.7.1 Classifiers for machine learning

ML, a type of artificial intelligence, has long been useful for identifying emotions. An ML classifier is constructed using a training dataset before being applied to the prediction of emotions. The recognition of emotions is the most common use of supervised ML algorithms. Researchers have looked at increasingly complex ML models to handle the demanding applications of today that involve emotion recognition. Moreover, a variety of ML techniques are applied to the multimodal emotion recognition problem, taking into account text, images, physiological inputs, and other modalities. ML algorithms ought to improve performance while managing a multitude of unique emotional characteristics.

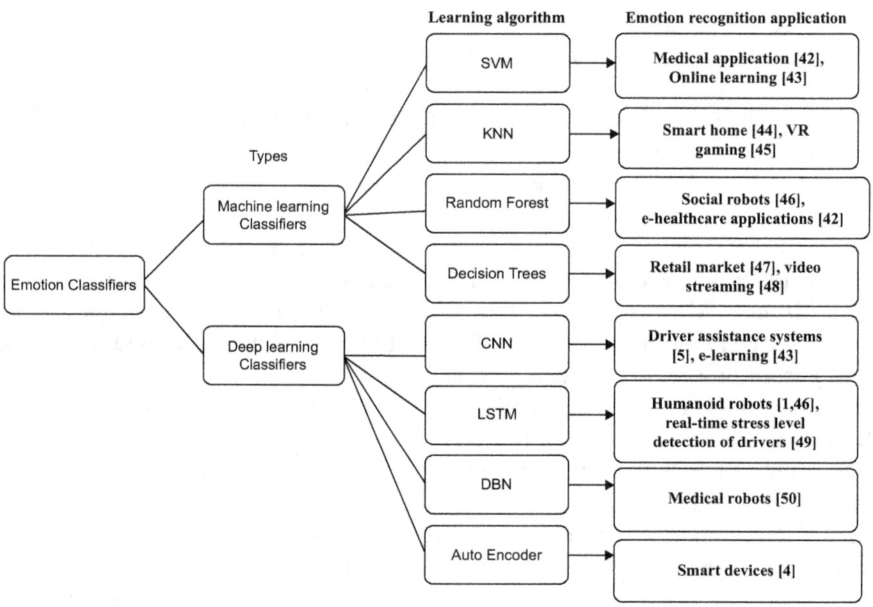

Figure 8.4 Emotion classifiers

8.7.2 Deep learning classifiers

The incapacity of conventional feature engineering and ML techniques to extract nonlinear patterns from multivariate time series data is their primary flaw [51].

Furthermore, in order to separate the most significant characteristics from a large amount of data, dimensionality reduction techniques are required. When given well-selected features that are retrieved using statistical variables like variance, entropy, and mean, ML models perform well [51]. The dimensionality of features increases the computational complexity of feature extraction. It's hard to pinpoint the perfect blend of attributes in general. To address the challenge of extracting pertinent characteristics from time series data, some researchers have focused on DL techniques. The time-consuming manual feature development procedure for ML. Almost every area of computing has significantly improved as a result of DL technologies like CNN, RNN, and autoencoder, especially computer vision, machine translation, and audio recognition. Recent work has utilized DL models, which offer high-level data abstraction, to develop reconfigurable frameworks for emotion recognition. Key variables are extracted from high-level data representations using deep neural networks [5].

DL techniques are useful because they can effectively handle unprocessed data and extract relevant features. Each nonlinear change in the network creates a hierarchical data representation structure, and the network is fed standard statistical samples.

8.7.3 Classifiers using machine learning versus deep learning

We looked at the detection accuracy of six ML and DL emotion classifiers from different research articles that used the DEAP dataset for emotion categorization. The DEAP dataset was used to test detection accuracy because it was the industry standard for multimodal emotion recognition in more than half of the reviewed studies. CNN, RNN, and AE are the three DL algorithms that work best together to create the finest DL model. Out of all the ML models we considered for our investigation, SVM had the best detection accuracy

8.8 Proposed model

The chapter presents a script that first detects the face in the image using openCV's Harr Classifiers. Initially, faces were identified using the frontal face cascade classifier. If a face was found, we clipped and saved the region of interest. Next, we locate our region of interest—in this example, the eyes—using the eye pair classifier, crop it, and save the freshly cropped image. Figure 8.4 depicts the cropping procedure.

8.8.1 Splitfed

The server-side software implementation is called splitfed. By dividing the total number of predictions by the total number of accurate forecasts, the forecast's accuracy is determined. Federated averaging is performed using the FedAvg function as a starting point. Next, splitfed creates n net model servers and one net server given n as the number of clients. The last functions to be defined are the server-side training and testing ones. The capability for client-side testing and training is then specified by the client-side software. Preprocessing entails reading a dataset, dividing it into train and test sets, and then converting each set into the necessary custom data class type. The testing and training processes then begin. We repeat the procedure according to the number of epochs (Figure 8.5).

8.8.2 FedSL

FedSL initializes the random seed along with several other parameters: number of rounds, number of local epochs, learning rate, number of clients, batch size, number of classes, and frac, which indicates the percentage of clients that will participate in the process. The dataset is then loaded, read, and divided into train and test sets. The dataset is subsequently shared among users only when indexes have been added. The model is then generated by a first and second part of an RNN. The testing and training processes then begin. By increasing the number of epochs and decreasing the number of clients for each epoch, we repeat the process. We can see the FedSL workflow in Figure 8.6.

8.8.3 PyVertical

The SplitNN class, which includes the models, optimizers, data, and distant tensors, is first created using PyVertical. Additionally, the class offers functions for gradient

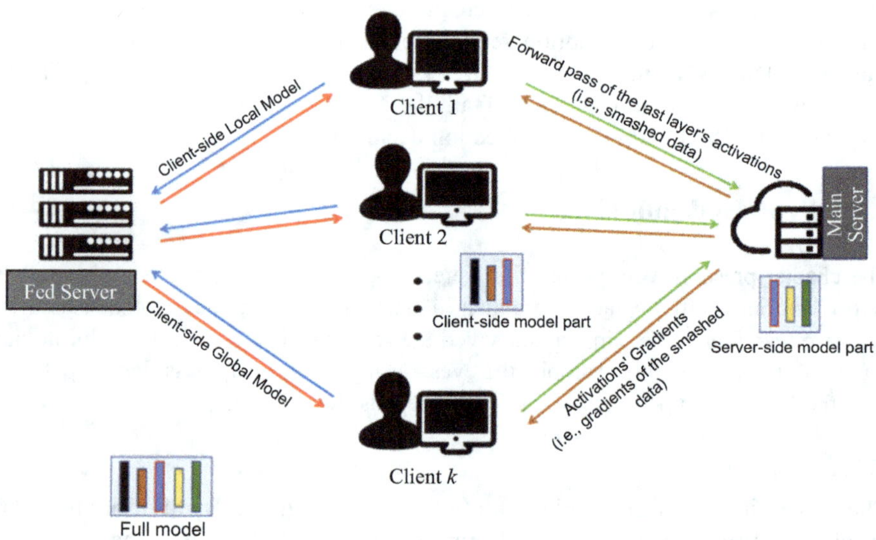

Figure 8.5 Workflow of splitfed

propagation backward and forward in data. Next, we compute the Private Set Intersection between the two encrypted sets, which ensures that neither set is disclosed to the other set save for the items in the intersection (i.e., the ids in this example). Next, PyVertical makes sure that the data is sorted by ids correctly. Next Two virtual workers, one holding the labels and the other the data, make up the model that is shown (Figure 8.7).

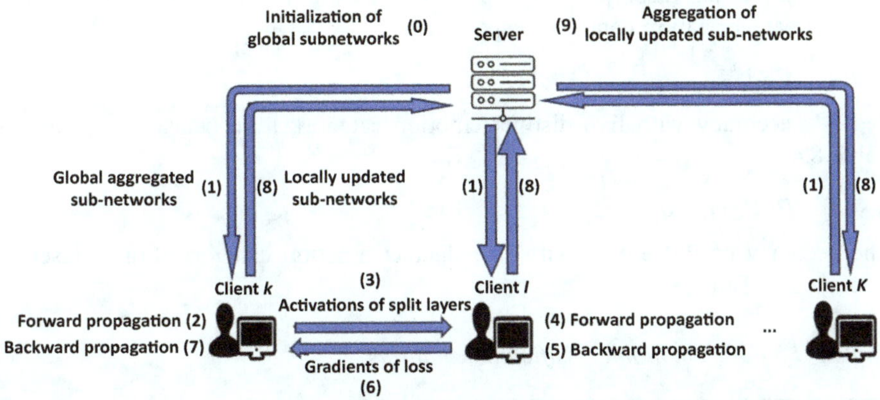

Figure 8.6 Workflow of FedSL

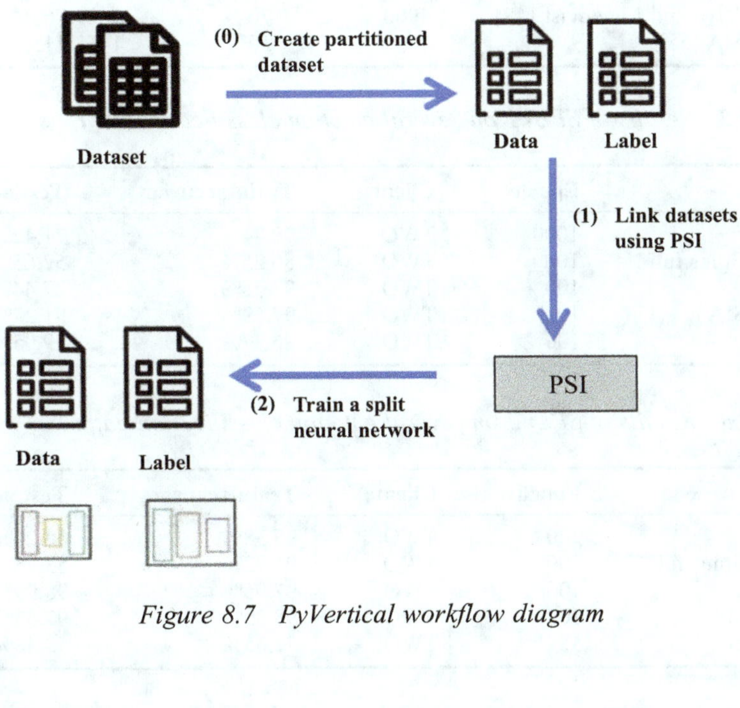

Figure 8.7 PyVertical workflow diagram

8.9 Results

8.9.1 Datasets

Five different datasets were compared and analyzed. The many datasets that were used are listed in Table 8.1.

8.9.2 SplitFed

Tables 8.2 and 8.3 display the accuracy of SFLV1 and SFLV2 using five distinct emotion categorization datasets, respectively.

8.9.3 FedSL

FedSL's accuracy with five distinct emotion categorization datasets is shown in Table 8.4.

8.9.4 PyVertical

The accuracy of PyVertical with five distinct emotion categorization datasets is displayed in Table 8.5.

Table 8.1 Face emotion classification datasets

Dataset name	Images	Subjects	Image	Date
FER2013plus [52]	30,000	30,000	48×48	2016
AffectNet [53]	1,000,000	450,000	48×48	2017
CKplus [54]	980	123	48×48	2010
KDEF Calvo and Lundqvist [48]	4900	70	562×762	1998
ouluCASIA [55]	1400	30	145×115	2011

Table 8.2 Results of SFLV1 on several emotion classification datasets

Dataset	Epochs	Clients	Train accuracy	Test accuracy
KDEF	1000	TWO	98.74%	71.42%
FER2013plus full	100	TWO	86.25%	52.65%
CKplus	100	TWO	97.58%	27.44%
ouluCASIA	100	TWO	97.59%	91.27%
AffectNet	100	TWO	86.77%	59.95%

Table 8.3 Results of SFLV2 on several emotion classification datasets

Dataset	Epochs	Clients	Train accuracy	Test accuracy
KDEF	100	TWO	97.38%	65.29%
FER2013plus full	100	TWO	91.72%	53.75%
CKplus	100	TWO	97.95%	98.99%
ouluCASIA	100	TWO	98.85%	97.53%
AffectNet	35	TWO	71.65%	53.48%

8.9.5 Comparison graph

Figure 8.8 presents a bar chart that compares the training accuracy of FedSL, PyVertical, normal centralized learning, federated learning, and split learning using five cropped emotion classification datasets with two users. They are all alike.

Table 8.4 The results of FedSL on several emotion classification datasets

Dataset	Epochs	Clients	Train accuracy	Test accuracy
FER2013plus full	20	TWO	35.36%	37.51%
CKplus	40	TWO	26.85%	27.45%
ouluCASIA	40	TWO	17.795%	15.69%
AffectNet	20	TWO	17.794%	17.65%

Table 8.5 Results of PyVertical on several emotion classification datasets

Dataset	Epochs	Clients	Train accuracy	Train loss
KDEF	1000	TWO	96.51%	0.199
FER2013plus Full	1000	TWO	98.98%	0.008
CKplus	1000	TWO	99.99%	0.008
AffectNet	895	TWO	99.93%	0.016
ouluCASIA	1000	TWO	91.76%	0.286

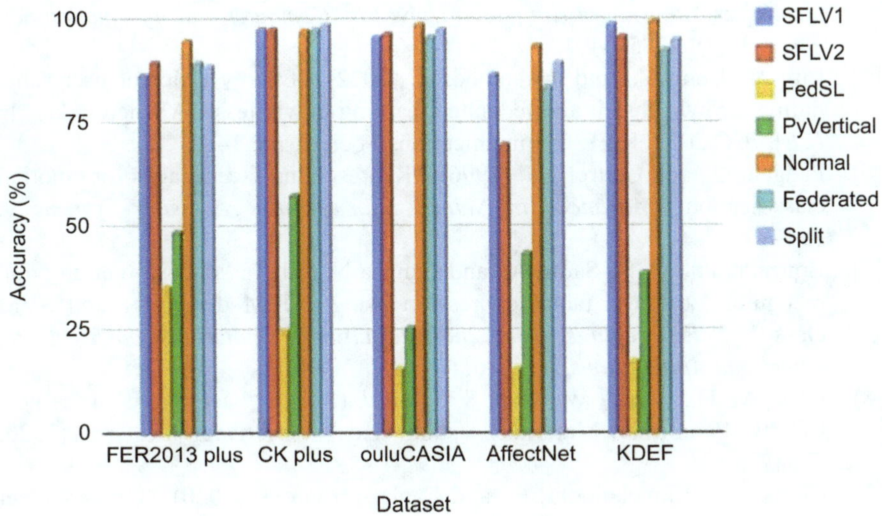

Figure 8.8 Comparison graph

8.10 Conclusion

This chapter presents a Face Emotion classifier that can identify an emotion from a cropped image that shows only the upper portion of the face. Split federated learning is a unique approach that combines federated learning and split learning to decrease federated learning computation time and circumvent federated learning resource limits while preserving user face privacy in the construction of the classifier. The three proposed split federated learning algorithms—FedSL, PyVertical, and SplitFed Versions 1 and 2—are examined and contrasted. While FedSL's performance was less striking, with relatively low accuracies of up to 35%, the classifier that used SplitFed and PyVertical fared well overall, with very high accuracies of up to 99% on some datasets.

References

[1] Baltrušaitis, T., Ahuja, C., and Morency, L. P. (2018). "Multimodal machine learning: A survey and taxonomy." *IEEE Transactions on Pattern Analysis and Machine Intelligence*, 41(2): 423–443.

[2] Abdullah, S. M. S. A., Ameen, S. Y. A., Sadeeq, M. A. M., and Zeebaree, S. (2021). "Multimodal emotion recognition using deep learning." *Journal of Applied Science and Technology Trends*, 2(2): 52–58.

[3] Alswaidan, N., and El Bachir Menai, M. (2020). "A survey of state-of-the-art approaches for emotion recognition in text." *Knowledge and Information Systems*, 62(8): 2937–2987.

[4] Dadebayev, D., Goh, W. W., and Tan, E. (2021). "EEG-based emotion recognition: Review of commercial EEG devices and machine learning techniques." *Journal of King Saud University-Computer and Information Sciences*, 34(7): 4385–4401.

[5] Liu, X., Deng Y., and Mahmoodi T. (2022). "Energy efficient user scheduling for hybrid split and federated learning in wireless UAV networks." In *IEEE ICC 2022*. IEEE Communications Society, pp. 1–6.

[6] Padgett, C., and Cottrell, G. (1996). "Representing face images for emotion classification." *Advances in Neural Information Processing Systems* 9: 894–900

[7] Bouzakraoui, M. S., Sadiq A., and Enneya N. (2017). "A customer emotion recognition through facial expression using POEM descriptor and SVM classifier." *Proceedings of the 2nd International Conference on Big Data, Cloud and Applications*, pp. 1–6.

[8] Chua, M. H., Cheng, W., Goh, S. S., *et al.* (2020). "Face masks in the new COVID-19 normal: Materials, testing, and perspectives." *Research* 2020: 7286735.

[9] Cimtay, Y., Ekmekcioglu, E., and Caglar-Ozhan, S. (2020). "Cross-subject multimodal emotion recognition based on hybrid fusion." *IEEE Access Practical Innovations Open Solutions*, 8: 168865–168878.

[10] Song, B. C., and Kim, D. H. (2021). "Hidden emotion detection using multimodal signals." *CHI Conference on Human Factors in Computing Systems Extended Abstracts* (pp. 1–7).

[11] Ahmed, N., Al Aghbari, Z., and Girija, S. (2023). "A systematic survey on multimodal emotion recognition using learning algorithms." *Intelligent Systems with Applications*, 17: 200171.

[12] Povolny, F., Matejka, P., Hradis, M., *et al.* (2016). "Multimodal emotion recognition for AVEC 2016 challenge." In *Proceedings of the 6th International Workshop on Audio/Visual Emotion Challenge* (pp. 75–82).

[13] Poria, S., Cambria, E., and Gelbukh, A. (2016). "Aspect extraction for opinion mining with a deep convolutional neural network." *Knowledge-Based Systems*, 108: 42–49.

[14] Lu, Y., Zheng, W.-L., Li, B., and Lu, B.-L. (2015). "Combining eye movements and EEG to enhance emotion recognition." *Proceedings of the Twenty-Fourth International Joint Conference on Artificial Intelligence* (pp. 1170–1176).

[15] Cen, L., Wu, F., Yu, Z. L., and Hu, F. (2016). "A real-time speech emotion recognition system and its application in online learning." In Tettegah, S. Y. and Gartmeier, M. (eds), *Emotions, Technology, Design, and Learning* (pp. 27–46). Amsterdam: Academic Press.

[16] Wei, K., Zhang, L., Guo, Y., and Jiang, X. (2020). "Health monitoring based on Internet of Medical Things: architecture, enabling technologies, and applications." *IEEE Access*, 8, 27468–27478.

[17] Du, G., Long, S., and Yuan, H. (2020). "Non-contact emotion recognition combining heart rate and facial expression for interactive gaming environments." *IEEE Access Practical Innovations Open Solutions*, 8: 11896–11906.

[18] Spezialetti, Matteo, Giuseppe Placidi, and Silvia Rossi. (2020). "Emotion recognition for human–robot interaction: Recent advances and future perspectives." *Frontiers in Robotics and AI*, 7: 532279.

[19] Imani, M., and Montazer, G. A. (2019). "A survey of emotion recognition methods with emphasis on E-Learning environments." *Journal of Network and Computer Applications*, 147: 102423.

[20] Sharma, G., and Dhall, A. (2021). "A survey on automatic multimodal emotion recognition in the wild." In Phillips-Wren, G., Esposito, A., and Jain, L.C. (eds.) (eds), *Advances in Data Science: Methodologies and Applications* (pp. 35–64). Cham: Springer.

[21] Saheb Jam, G., Rhim, J., and Lim, A. (2021). "Developing a data-driven categorical taxonomy of emotional expressions in real world human robot interactions." In Companion of the 2021 ACM/IEEE International Conference on Human-Robot Interaction (pp. 479–483).

[22] Lim, J. Z., Mountstephens, J., and Teo. J. (2020). "Emotion recognition using eye-tracking: taxonomy, review and current challenges." *Sensors*, 20 (8): 2384.

[23] Shirke, B., Wong, J., Libut, J. C., George, K., & Oh, S. J. (2020). "Brain-IoT based emotion recognition system." In *10th Annual Computing and Communication Workshop and Conference (CCWC)* (pp. 991–995). IEEE.

[24] Hipson, W. E., and Mohammad, S. M. (2021). "Emotion dynamics in movie dialogues." *PLoS One*, 16(9): e0256153.

[25] Subasi, A., Tuncer, T., Dogan, S., Tanko, D., and Sakoglu, U. (2021). "EEG-based emotion recognition using tunable Q wavelet transform and rotation forest ensemble classifier." *Biomedical Signal Processing and Control*, 68: 102648.

[26] Bosch, E., Klosterkamp, M., Guevara, A., Kaethner, D., Bendixen, A., and Ihme, K. (2022). "Multimodal estimation of frustrative driving situations using a latent variable model." *13th IEEE International Conference on Cognitive Infocommunications (CogInfoCom)* (pp. 11–16). IEEE.

[27] Suja, P., Thomas, S. M., Tripathi, S., and Madan, V. K. (2016). "Emotion recognition from images under varying illumination conditions." In Balas, V., Jain, L., and Kovačević, B. (eds) (eds.), *Soft Computing Applications* (pp. 913–921). Cham: Springer.

[28] Bosch, E. J., Kathner, D., and Drewitz, U. (2021). "Evidence for individual-specific expressions of frustration." *Proceedings of the European Society for Cognitive and Affective Neuroscience*. https://elib.dlr.de/143073/.

[29] Marsella, S., Gratch, J., and Petta, P. (2010). "Computational models of emotion." In Scherer, K.R., Banziger, T., and Roesch, E. (eds.) (eds), *A Blueprint for Affective Computing: A Sourcebook and Manual* (pp. 21–46). New York: Oxford University Press.

[30] Gjoreski, H., Mavridou, II., Fatoorechi, M., *et al.* (2021). "emteqPRO: Face-mounted mask for emotion recognition and affective computing." In *Proceedings of the Adjunct ACM International Joint Conference on Pervasive and Ubiquitous Computing and Proceedings of the ACM International Symposium on Wearable Computers* (pp. 23–25).

[31] Nakisa, B., Rastgoo, M. N., Rakotonirainy, A., Maire, F., and Chandran, V. (2018). "Long short term memory hyperparameter optimization for a neural network based emotion recognition framework." *IEEE Access*, *6*, 49325–49338.

[32] Majid Mehmood, R., Du, R., and Lee, H. J. (2017). "Optimal feature selection and deep learning ensembles method for emotion recognition from human brain EEG sensors." *IEEE Access*, 5: 14797–14806.

[33] Tang, H., Liu, W., Zheng, W. L., and Lu, B. L. (2017). "Multimodal emotion recognition using deep neural networks." In Liu, D., Xie, S., Li, Y., Zhao, D., and El-Alfy, E. S. (eds.), *Neural Information Processing. ICONIP 2017* (pp. 811–819). Cham: Springer.

[34] Hassan, M. M., Alam, M. D. G. R., Uddin, M. D. Z., Huda, S., Almogren, A., and Fortino, G. (2019). "Human emotion recognition using deep belief network architecture." *Information Fusion*, 51: 10–18.

[35] Shukla, J., Barreda-Ángeles, M., Oliver, J., Nandi, G. C., and Puig, D. (2021). "Feature extraction and selection for emotion recognition from

electrodermal activity." *IEEE Transactions on Affective Computing*, 12(4): 857–869.

[36] Ragot, M., Martin, N., Em, S., Pallamin, N., and Diverrez, J. M. (2018). "Emotion recognition using physiological signals: Laboratory vs. wearable sensors." In Ahram, T., and Falcão, C. (eds.), *Advances in Human Factors in Wearable Technologies and Game Design. AHFE 2017* (pp. 15–22). Cham: Springer.

[37] Quiroz, J. C., Geangu, E., and Yong, M. H. (2018). "Emotion recognition using smart watch sensor data: Mixed-design study." *JMIR Mental Health*, 5 (3): e10153.

[38] Vergara, D., Extremera, J., Rubio, M. P., and Dávila, L. P. (2019). "Meaningful learning through virtual reality learning environments: A case study in materials engineering." *Applied Sciences*, 9(21): 4625.

[39] Marín-Morales, J., Higuera-Trujillo, J. L., Greco, A., *et al.* (2018). "Affective computing in virtual reality: emotion recognition from brain and heartbeat dynamics using wearable sensors." *Scientific Reports*, 8(1): 13657.

[40] Koelstra, S., Muhl, C., Soleymani, M., *et al.* (2011). "DEAP: A database for emotion analysis; using physiological signals." *IEEE Transactions on Affective Computing*, 3(1): 18–31.

[41] Goodfellow, I. J., Erhan, D., Carrier, P. L., *et al.* (2013). Challenges in representation learning: A report on three machine learning contests. In *Proceedings of the International Conference on Neural Information Processing* (pp. 117–124). Berlin: Springer.

[42] Elshabrawy, K. M., Alfares M. M., and Salem M. A-M. (2022). "Ensemble federated learning for non-II D COVID-19 detection." In *2022 5th International Conference on Computing and Informatics (ICCI)*. Piscataway, NJ: IEEE, pp. 57–63.

[43] Gao, Y., Kim, M., Abuadbba, S., *et al.* (2020). "End-to-end evaluation of federated learning and split learning for internet of things." *International Symposium on Reliable Distributed Systems (SRDS)*, Shanghai, China, pp. 91–100.

[44] Bota, P. J., Wang, C., Fred, A. L. N., and Da Silva, H. P. (2019). "A review, current challenges, and future possibilities on emotion recognition using machine learning and physiological signals." *IEEE Access Practical Innovations Open Solutions*, 7: 140990–141020.

[45] Joshi, P., Thapa, C., Camtepe, S., Hasanuzzamana, M., Scully, T., and Afli, H. (2021). "Splitfed learning without client-side synchronization: Analyzing client-side split network portion size to overall performance." arXiv:2109.09246, https://doi.org/10.48550/arXiv.2109.09246.

[46] Pathak, R., and Wainwright M. J. (2020). "FedSplit: An algorithmic framework for fast federated optimization." *Advances in Neural Information Processing Systems*, 33: 7057–7066.

[47] Ads, O. S., Alfares M. M., and Salem M. A-M. (2021). "Multilimb split learning for tumor classification on vertically distributed data." In *2021 10th International Conference on Intelligent Computing and Information Systems (ICICIS)*. Piscataway, NJ: IEEE, pp. 88–92.

[48] Calvo, M. G., and Lundqvist D. (2008). "Facial expressions of emotion (KDEF): Identification under different display-duration conditions." *Behavior Research Methods*, 40(1): 109–115.

[49] Elor, A., Powell, M., Mahmoodi, E., Hawthorne, N., Teodorescu, M., and Kurniawan, S. (2020). "On shooting stars: Comparing cave and HMD immersive virtual reality exergaming for adults with mixed ability." *ACM Transactions on Computing for Healthcare*, 1(4): 1–22.

[50] Guo, Y., Yang, Y., Liu, Y., *et al.* (2021). "Development status and multilevel classification strategy of medical robots." *Electronics*, 10(11): 1278.

[51] Aiswaryadevi, V. J., Priyanka, G., Sathya Bama, S., *et al.* (2021). "Smart IoT multimodal emotion recognition system using deep learning networks." In Manoharan, K. G., Nehru, J. A., and Balasubramanian, S. (eds), *Artificial Intelligence and IoT Smart Convergence for Eco Friendly Topography* (pp. 3–19) Singapore: Springer.

[52] Zahara, L., Musa, P., Wibowo, E. P., Karim, I., and Musa, S. B. (2020). "The facial emotion recognition (FER-2013) dataset for prediction system of micro-expressions face using the convolutional neural network (CNN) algorithm based Raspberry Pi." In *2020 5th International Conference on Informatics and Computing (ICIC)*. Piscataway, NJ: IEEE, pp. 1–9.

[53] Mollahosseini, A., Hasani B., and Mahoor, M. H (2017). "Affectnet: A database for facial expression, valence, and arousal computing in the wild." *IEEE Transactions on Affective Computing*, 10(1): 18–31.

[54] Adhikary, A., and Bhandari N. (2021). "PosEmotion-combining realtime 2D body pose estimation and facial emotion recognition to analyze human behavior." In *2021 26th International Conference on Automation and Computing (ICAC)*. Piscataway, NJ: IEEE.

[55] Zhao, G., Huang, X., Taini, M., Li, S. Z., and PietikäInen, M. (2011). "Facial expression recognition from near-infrared videos." *Image and Vision Computing*, 29(9): 607–619.

Chapter 9

Split federated learning-based educational data analysis

Vamshi Krishna B.[1], Geetabai S. Hukkeri[1], Raguru Jaya Krishna[1] and T. Gopalakrishnan[2]

This study provides an up-to-date evaluation of predictive models used in forecasting student performance and learning outcomes in the educational sector. It examines modern machine learning and deep learning models for predicting college dropouts, utilising cost-effective large-scale administrative data as an early warning system. Additionally, the study investigates strategies for ensuring data privacy through federated and split federated learning techniques in forecasting dropout rates across a diverse student population. Furthermore, it addresses the importance of safeguarding data privacy, including student performance, user profiles and personalised learning paths, within the educational sector. The chapter encompasses big data analysis in education, exploring patterns, highlighting research themes, identifying limitations and proposing future directions. It also delves into the application of splitfed learning techniques to the prominent learning analytics challenge of predicting student dropouts, thereby contributing to both predictive accuracy and data privacy in educational settings.

Keywords: Machine learning; Deep learning; Federated learning; Split federated learning

9.1 Introduction

University administrators aim to lower high college dropout rates, which requires efforts to study and address the causes of student dropout. The advent of Big Data has given rise to educational process mining, allowing for new ways to organise and convey data. These findings assist researchers and administrators in assessing the efficacy of various data sources for developing early warning systems and

[1]Department of Computer Science and Engineering, Manipal Institute of Technology Bengaluru, Manipal Academy of Higher Education, India
[2]Department of Information Technology, Manipal Institute of Technology Bengaluru, Manipal Academy of Higher Education, India

making informed decisions aligned with specific policy objectives, thereby contributing to the long-term goal of preventing college dropout in today's post-secondary education institutions.

Machine learning (ML), a pivotal subset within artificial intelligence (AI), has risen as the transformative force reshaping the landscape of knowledge discovering processes [1]. AI technology relies heavily on ML. Researchers at all levels often regard ML as an information gathering process with a definite goal. It concentrates not only on gaining knowledge but also on enhancing ML abilities.

ML is the study of how machines may learn new abilities and knowledge by harnessing existing information and insights.

Nowadays in educational sector, ML and big data analytics play a major role for gathering data from various sources to analyse student performance, user profiles and personalised learning paths can take advantage of federated learning (FL) because it allows protecting sensitive data in the original source. The authors for [2] done sophisticated analysis to apply ML methods to predict student achievement, customised learning pathways and focus on Science, Technology, Engineering and Mathematics (STEM) education. In this article, authors proposed a complete directive to stakeholders with respect to data privacy in education field [3]. Educational data frequently contain several categorical factors, which influence the efficacy of prediction models for predicting student achievement. Learning management systems are increasingly being used to manage instructional resources, oversee activities and monitor student behaviour to ensure long-term education [4]. This study investigates the elements that influence undergraduate academic achievement, demonstrating a link between students' social and academic qualities and success. Addressing dropouts and success rates in higher education, the study [5] emphasises the necessity of bridge courses in cooperative study programmes and proposes ways to mitigate factors that contribute to low academic performance. In [6,7], the authors discussed the students' path in achieving a graduate degree or a dropout. Furthermore, the paper [8] assesses the predictive efficacy of ML models across various student groups and suggests a clustering-based MOOC dropout detection method as well as an early prediction model based on deep learning. The analysis focuses on visualising students' routes through a study programme based on their 'dropout' or 'graduate' status, as well as their degree programme level is studied in [9].

In ML model, data is collected centrally, and the model is also trained using the aggregated data in a single machine. In this current era, data privacy and protection are an utmost concern for various organisations. FL is a subfield of ML; in this approach, data is kept local to the servers and multiple devices are involved in training the model. FL is decentralised technique to train the model and acquire the required weights depending on the data provided locally in the systems from multiple devices.

In FL, the training process is performed across various devices like smartphones, laptops or servers in a distributed environment. For the data which is available locally to each device, a local model is trained and updated weights are sent to the central location for further processing. At the central location or server, updated weights are received from multiple devices and global model is created by processing the aggregated weights received from multiple sources. The global model updates are sent back to various devices for additional enhancement. The major advantage of FL is data remains locally on a device and data privacy risks associated with centralised data storage can be avoided.

Deep learning has got its attention in the various research domains and in industry practices during the current decade, which revolutionised the field of AI and ML through achieving versatile results in most of the complex applications and domains such as text, image, speech recognition and recommendation systems [10].

The important characteristics feature of any deep learning application is to maximise the utilisation of deep neural network concepts, which basically have more than one or *n*-multiple layers of interconnected computing nodes. These interconnected networks help in learning complex representations hidden inside the data repositories by discovering various hierarchical patterns and features in the available data. The objective of any deep learning algorithm is to learn the data representations automatically and enhance the feature set to obtain better results.

FL, also known as collaborative learning, is a decentralised method of training ML models that eliminate the need to transport data between client devices and worldwide servers. Instead, the model is trained locally on raw data kept on edge devices, which improves data privacy while maintaining advantages in terms of privacy, security and access to diverse data. In the area of higher education, data is quite sensitive and has restrictions to access it. Through FL approach it is possible to provide a safe model training by providing access to education data in local environment without any necessity for centralisation.

The FL approach aids institutional collaboration, allowing models to learn from a diversity of datasets in a protected environment. However, the distributed nature of FL presents obstacles, such as heterogeneity in systems, data and device capabilities, which must be addressed to increase the accuracy of shared models. This study focuses on forecasting student dropout in educational institutions using FL methodologies.

9.2 Related work

ML traces its roots back to the 1950s, gaining recognition as a distinct field in the 1990s [11]. Despite its longstanding history, education research in ML remains nascent, primarily due to its recent inclusion in university curricula [12]. According to Georgiopoulos *et al.* [13], ML courses have been available at academic institutions for decades, though formalised education in the subject is relatively new. Positioned as a subset of computer science, ML courses typically attract students majoring in computer science or data science [14]. Today, ML is integrated into university curricula, spanning various disciplines beyond computer science, and typically offered in the second and third years of study [15].

The adoption of ML heralds a shift in problem-solving methodologies, transitioning from traditional analytical approaches to robust data-driven methodologies.

This transformation empowers software programs to construct models from training data and extrapolate insights from new datasets [16]. Despite the proliferation of ML applications, literature on teaching methodologies remains limited. Ho and Scadding [17] highlight a common challenge faced by instructors: teaching technological-oriented subjects to students with varying levels of interest in technology. Typically, teaching ML revolves around two core components: imparting knowledge of ML techniques and their practical application. This pedagogical approach often involves introducing students to various ML techniques and facilitating hands-on experience with specific software packages using sample datasets.

As ML spreads across domains [18–20], it is possible to transform academic research and enactment becomes clearer [21,22]. A wide range of ML approaches, including, unsupervised, supervised and reinforcement learning, are used to address a variety of educational issues and opportunities [23–25]. These applications include intelligent systems for tutoring, adaptive learning channels, education analytics and performance of pupils' prediction, among other things [26].

In [27], Tsai *et al.* investigate that the role of big data and deep learning techniques is to help universities or higher education institutes (HEI) to precisely understand the various student backgrounds. This paper also discusses the probability to predict student dropout using a statistical learning method based on deep neural networks.

Amin Al Ka'bi *et al.* [28] proposed an AI and deep learning-based model to improve the cognitive capability of students of higher education in comparison with other techniques. The results show enhanced performance compared to other standard reference models. Abdallah Moubayed *et al.* [29] proposed a predictive model that can predict the students' grade at the end of the graduation in the ongoing duration itself. This will help the students to improve by identifying their reasons for deviation in obtaining the final grades. This predictive approach uses deep learning models to classify students based on numerous academic and non-academic tendencies and forecast their final grades at the halfway point of the course duration.

Mustafa Yağcı [30] proposed a model depending on the information available during midterm exam scores and grades and the authors were able to predict the grades of undergraduate students in their final examinations. This framework predicts the use of major parameters such as faculty, midterm exam grades and department information.

In [31], the authors have discussed a design elements in FL and highlighted the significance of autonomy and heterogeneity in real-world FL design. In [31], the authors also presented entire classification and possibilities through a comprehensive framework for classifying FL that takes six factors into account. They also found potential in the industry such as the incorporation of new platforms like blockchain and the requirement for more standards.

In [32], the authors have discussed a detailed analysis of FL like comprehensive analysis of FL design, architecture and implementation, a comparison between distributed and centralised ML systems, novel taxonomies and classification systems like creation of a new classification scheme for research disciplines and FL subjects, taxonomies encompassing important technological elements and new fields such as resource management, privacy, security, application domains and system architecture.

In [33], the authors have provided a detailed understanding of the technological foundations, factors and characterisation of FL. In [33], the authors have also given a comprehensive analysis of privacy and modern applications with respect to a wide range of topics in its in-depth assessment of Florida privacy considerations. The authors also focused on the most popular applications now being used in industry areas (data science, healthcare, education, etc.) and across technologies (AI, IoT, blockchain, etc.).

9.3 Research background

9.3.1 Machine learning

ML encompasses three primary areas of investigation:

1. Fundamental research centers on the theory and methodologies of ML, drawing parallels to human learning processes.
2. Cognitive science is strongly related to the growth of ML, establishing a cognitive physiology model by simulating human learning processes.
3. Applied research strives to enhance current methodologies. ML can integrate novel techniques rooted in its distinctive characteristics, emulate human learning strategies or amalgamate both to innovate new methodologies.

ML contributes significantly to education assessment of data by offering insights, predictions and tailored learning experiences. Some of the areas ML in education include:

- Predictive analytics: Using available past higher education data, ML algorithms can predict future performance of the students like average time to complete a course, dropout prediction, early identification of academically weak students which can help instructor as well as student to provide necessary support.
- Personalised learning: An individual learning path for each student can be recommended to students with the help of ML algorithms. With the availability of personalised learning paths and with the use of various educational platforms, different types of course resources, learning techniques and activities can be recommended to a student.
- Adaptive learning systems have been made possible with ML that dynamically adjusts content and difficulty to suit student progress. These systems can constantly check student performance and adapt educational materials to ensure the highest quality learning environment.
- ML for learning patterns: By analysing big data sets, machine-learning algorithms can identify patterns and trends in students' learning activities. Understanding students' interactions with instructional materials helps improve curriculum design, teaching methods and evaluation procedures aimed at enhancing the efficiency of the process of education.

- Natural language processing (NLP) in education: Analyse texts comprising student essays, forum discussions or instructional resources as NLP applications in education like sentiment analysis, summarisation or subject modelling among others can give insights concerning students' participation, comprehension and writing capability.
- Educational recommender systems utilise ML algorithms to suggest suitable learning resources, courses or materials based on learners' interests, objectives as well as prior behaviours. This way they support children as they look for new knowledge and remain engaged during their studies.

In summary, by proposing various levels of instructions and recommendations, learning experience of all levels of students can be improved by using ML in education data analysis.

9.3.2 Deep learning

To address prediction issues, deep learning models are introduced. These models are typically based on self-learned representations that closely follow the artificial neural networks (ANNs), which attempt to mimic the information processing in the human brain. Deep learning algorithms can be used to create a predictive model of academic achievement, which is adaptive and, at the same time, sensitive to diverse components that influence educational outcomes.

This chapter investigates the use of numerous deep learning algorithms and data encoding methods, with a focus on their function in fostering sustainable development in education. Objectives include:

- Evaluating and predicting student success.
- Recognising how deep learning approaches can be used to predict student success.

Deep learning architectures include well-known models such as convolutional neural networks (CNNs), deep belief networks (DBNs), long short-term memory networks (LSTMs), generative adversarial networks (GANs), recurrent neural networks (RNNs) and others. In the field of education, these models are used to identify student involvement, facilitate interactive learning methods, provide personalised learning environments (particularly in online platforms) and occasionally serve as assessment evaluation tools.

9.3.3 Federated learning

To address the issues related to ML in central training, AI models started to shift to a decentralised approach, and a new notion called 'federated learning' has developed. The FL paradigm involves a central service provider hosting a global ML model. This model is then transmitted to a set of customers or users via their local edge devices. The approach begins with training each client's data on their respective local devices, utilising the global model as a foundation. This leads in the establishment of a trained local model, with the client's data safely stored on their own edge device.

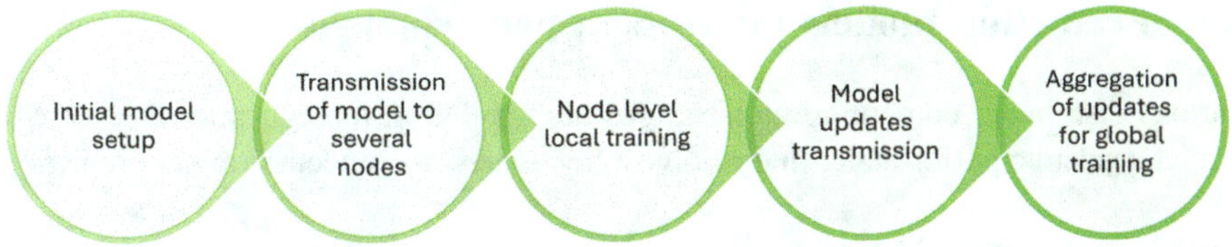

Figure 9.1 Federated learning model training phases

The updated parameters of the clients are sent back to the central server after local training is finished. The global model's parameters are then updated by averaging these values over the whole group of clients who are involved. This completes the first FL iteration or training round. After the global model update, a second training cycle is started by broadcasting the improved model to a freshly chosen pool of clients. In Figure 9.1, the training phases of FL are shown.

FL functions as a collaborative, iterative learning method. Because of its modular design, it creates an atmosphere that is favourable for private and safe collaborative ML, advancing models without jeopardising the privacy of specific client information.

FL in higher education: Accessibility towards ML-based technologies in everyday activities has seen a significant increase in the number of applications. However, there are always several challenges associated with extensive acceptance of ML in higher education.

Some of the major challenges of ML-based application of higher education are as follows:

- Data security and privacy: ML models need large volumes of data for accurate predictions and recommendations. Personal data protection cannot be compromised in the name of enhanced learning experience. There is a significant amount of shortage in providing data protection, during central data processing.
- Infrastructure issues: ML models need huge resources like CPUs and computation processing. Training and ratification of models without proper infrastructure and data might lead to improper information and in turn lead to flawed predictions.

9.3.4 Split federated learning

In distributed computing environments to address security and privacy concerns, a novel technology called split federated learning has emerged, which is quite like FL technology. Deep neural networks are partitioned into two sections like FL and split federated learning. Using these techniques, data security can be improved even when data is shared among multiple devices or clients for training. In higher education with the use of split federated learning without negotiating privacy concerns, models can be trained. This approach can help educational sectors have better models to identify trends and patterns in student data, thus leading to improved student performance and personalised learning paths.

9.4 Prediction models for student performance

Various distributed computing architectures are applied in the educational sectors to predict academic performance and resolve a few issues like student dropout prediction.

9.4.1 Machine learning

ML approaches have proven invaluable in forecasting learner's academic progress. Traditionally, the conventional strategy was to collect all data on a central server for model training. Nonetheless, while this method is simple, it raises privacy concerns because it limits access to a large amount of useful but sensitive data. Table 9.1 shows a brief summary of ML models used for student performance prediction systems.

9.4.2 Deep learning

Deep learning techniques have played a critical role in predicting student's academic performance. Ange Tato *et al.* [39] addressed this issue by proposing a hybrid architecture which combines the layered approach of deep neural network architectures – especially LSTM and CNN. This combined architecture provides the use of attention mechanism which will infuse expert knowledge into the deep neural network. This combined hybrid model was tested in means of two real-time cases: the first one involves identifying the knowledge in an intelligent tutoring system and the second method focuses on prediction of socio-moral reasoning.

Table 9.1 A brief review of machine learning algorithms

Reference	Algorithms	Dataset	Attributes	Better algorithm
[34]	Naïve Bayes, SCM, ID3C-4.5.	'UCI Machinery Student Performance'	18	SVM
[35]	Naïve Bayes, KNN	'Gaza Strip 2015'	Gender, DOB, City, Fathers Job, Secondary School Name, Student Status, Specialisation	Naïve Bayes
[36]	Naïve Bayes (NB)	'Amrita School of Arts and Science, Mysure'	19	Naïve Bayes
[37]	Linear regression (LR) for deep learning and supervised learning	'Kaggle Dataset'	22	Linear regression for supervised learning
[38]	Decision tree (DT), naïve Bayes	'Industrial Engineering University, Islam, Indonesia'	1st sem GPA, MEDU, Attendance	Naïve Bayes

Figure 9.2 shows how this hybrid model has been integrated to achieve the expected outcome of students' knowledge tracing. Table 9.2 shows the brief summary of deep learning models used for student performance prediction systems.

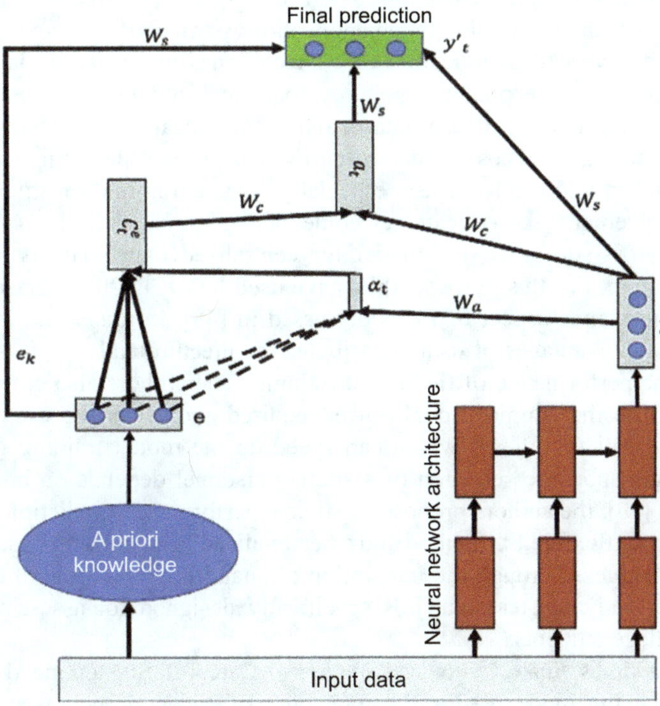

Figure 9.2 Global attentional hybrid model by Ange Tato [39]

Table 9.2 A brief review of deep learning algorithms

Reference	Algorithms	Dataset	Attributes
[40]	Deep neural network	Multidisciplinary university	Sixteen academic units (faculty/institute/college)
[41]	mBERT +FCNN	Hohai University Real-time dataset of 236 numbers	Various subjects related to ideological and political education in consideration of the subject context
[42]	ReLU and Sigmoid, Functions	Open University Learning Analytics dataset	Courses, students and their interactions with the virtual learning environment (VLE)
[43]	CNN-LSTM deep learning model	Blackboard LMS Dataset	Students' KPIs
[44]	Autoencoder-Deep neural networks-K-Means and SMOTE	Open University Learning Analytics (OULA) dataset	Logs of students' online learning activities

9.4.3 Federated learning

FL techniques have played a critical role in predicting student's academic performance.

An overview of multi-agent reinforcement learning overview and its application in distributed environment is discussed in [45]. FL is introduced by Google [46] in 2015 for various mobile phone applications particularly for text prediction-based Gboard keyboard. Application of FL techniques in education data analysis is a perfect match. One potential solution to ethical and legal issues related to handling sensitive data is to use data analytics in education, especially learning analytics. Robust guarantees for data security and privacy protection are crucial for study conduct and result release, especially considering the sensitive nature of the data involvement like educational comes and personal data. In [47], authors proposed a cross-silo-based FL model for centralised data analysis scheme in education system. For this purpose, they have used KDDCup2015 dataset.

The performance aspect of FL is discussed in [48]. In this work, the authors have discussed advantages of using distributed architecture and how it is useful in improving the performance of the model training. The authors also conveyed that FL could reduce the computational power required and can make use of parallel training across all the clients which can speed up the model training greatly. In general, in any sport the selection of a sports personnel depends on his past performance. In [49], the authors proposed a student performance prediction technique by introducing a flexible FL method and a personalised FL algorithm based on deep learning techniques. Through the integration of quantitative motor skill evaluation methods, the study creates standards specifically designed for assessing athletic ability in college students.

In recent times, many renowned universities are offering online degree programmes and in this mode of education, prediction of student dropout is an important task. Tradition approaches are most of the time not suitable for online education programmes. In [50], the authors have discussed various ML models to predict student dropout rates from multiple online institutes. In this article, a predictive classifier is proposed using a deep learning-based algorithm. In [51], the authors have proposed a pattern aware FL-based dropout prediction model for MOOC platforms, which uses 1D CNN and bidirectional LSTM layer as an FL framework.

Due to increasing use of online learning platforms and availability of ample number of distracting applications like social networking applications, it is hard for learners to remain focused on their learning platforms. In this regard, in [52] authors proposed a novel privacy preserving architecture approach to track the utilisation of screen time for E-learning using FL technology. In their proposed model they captured screenshots for every minute, and it is analysed for either the screenshot in entertainment class or learning class.

Various distributed computing architecture are applied in the educational sectors to predict academic performance and resolve a few issues like student dropout prediction. In [49], the authors discussed and presented a DL-based model for student dropout prediction. In this model, they have used various parameters

Table 9.3 A brief review of federating learning systems

Author	Year	Scope
Banabilah *et al.* [33]	2022	A comprehensive view of FL fundamentals is provided and emphasised more on AI, IoT, blockchain, etc.
Duan *et al.* [52]	2022	This article presents the most recent advancements in FL and split learning about edge computing-based Internet of Things environments.
Li *et al.* [31]	2021	FL background, analysis of systems point of view, data privacy and protection provided a design factor for FL and solutions for some scenarios
Alazab *et al.* [53]	2021	Need of FL for various cyber security and cyber attacks were discussed
AbdulRahman *et al.* [32]	2020	A thorough analysis of FL, in-depth analysis of FL design, architecture and comparison between centralised and decentralised ML algorithms
Li *et al.* [46]	2020	Presented review of applications in FL

like discussions viewed, HTML pages completed, videos viewed, exams completed, self-assessments completed, posts viewed, discussions held, tasks performed, contacted tutor, hours spent and number of times visited courses from the dataset to train the DL model. In this model, a DL classifier-based prediction model was used to classify student performance. In this model, a bidirectional LSTM model is used to depict dependencies and patterns in successive data.

In [50], the authors proposed an FL pattern aware model for dropout prediction. In their proposed model, they have used a 1D CNN and bidirectional LSTM to analyse temporal dynamics of the dataset. Table 9.3 shows the summary of FL models used for student performance prediction systems.

9.4.4 Split federated learning

Split federated learning approaches are very crucial in predicting students' academic progress by putting more analysis on data in the educational industry. Although evolving split learning (SL) solutions reduce the computational complexity on the client side by taking part in the model architecture, a large amount of time delays are encountered in the SL-based approaches. In every global round backward gradient and forward activations cost for transmission between local client and central server needs to be improved.

In [54], a unique FL model was presented to predict student learning outcomes across many institutions while maintaining data privacy. The model follows a standard to protect dataset confidentiality and predicts student grades in four levels: low, good, average and drop. In [55], a framework based on federated transfer learning was suggested for student categorisation with privacy protection, combining domain adaption methods within FL limitations. Furthermore, in [56], a hybrid federated SL architecture was proposed to combine the efficiency and privacy advantages of both approaches. Han *et al.* [57] developed a new federated

Table 9.4 A brief review of split federated learning systems

Author	Year	Data set	Scope
Umer Farooq	2023	Kaggle datasets	The study includes optimising characteristics, testing ML models and evaluating their performance.
Bin Xu	2022	Datasets of Portuguese schools	A framework based on effectively handling federated transfer learning on diverse datasets.
Turina	2021	MNIST dataset	Hybrid federated split learning improves latency and memory consumption.
Han	2021	MNIST and FMNIST	Splitting the model to reduce latency and training large-scale models.
Qiang Duan	2022	Multiple datasets	Combining federated and split-fed learning approaches.
Chandra	2022	MNIST, FMNIST, CIFAR10	Privacy and robustness measures are based on differential privacy.

SL technique that updates client- and server-side models simultaneously using locally adapted SL training. Furthermore, Tan *et al.* [58] reviewed the most recent advances in FL and SL, giving a summary of cutting-edge technology for integrating these methodologies in an edge computing-based IoT context. Finally, in [59], splitfed learning was introduced to improve privacy via network splitting and differential private client-side model updates. In [60] authors have discussed about student dropout prediction using federated learning techniques. Table 9.4 shows the summary of the SL models used for student performance prediction systems.

9.5 Conclusion

This study provides an up-to-date evaluation of the most recent models used in predicting student performance and learning in the educational sector. It effectively investigates modern ML and deep learning models for predicting college dropouts, using cost-effective large-scale administrative data as a viable early warning system. Furthermore, the study investigates strategies for ensuring privacy through federated and split federated learning techniques in forecasting dropout rates across a large student population. Furthermore, predictive analytics technologies are used to address the different elements that influence students' academic success.

References

[1] Lin, P., Van Brummelen, J., Lukin, G., Williams, R., and Breazeal, C. (2020). Designing a conversational agent for children to explore machine learning concepts. In *Proceedings of the AAAI Conference on Artificial Intelligence*, 34(9), 13381–13388.

[2] Abdrakhmanov, R., Zhaxanova, A., Karatayeva, M., Niyazova, G. Z., Berkimbayev, K., and Tuimebayev, A. (2024). Development of a framework

for predicting students' academic performance in STEM education using machine learning methods. *(IJACSA) International Journal of Advanced Computer Science and Applications*, 15(1), 28–46.

[3] Ouahi, M., Khoulji, S., and Kerkeb, M. L. (2024). Advancing sustainable learning environments: A literature review on data encoding techniques for student performance prediction using deep learning models in education. *E3S Web of Conferences*, vol. 477, 00074. https://doi.org/10.1051/e3sconf/202447700074.

[4] Salari, M., Radfar, R., and Faghihi. M. (2024). Predicting students' performance using machine learning algorithms and educational data mining (a case study of Shahed University). *Journal of Business Intelligence Management Studies*, 12(47), 315–366.

[5] Wild, S., Rahn, S., and Meyer, T. (2024). Participation in bridging courses and dropouts among cooperative education students in engineering. *International Journal of Education in Mathematics, Science, and Technology (IJEMST)*, 12(2), 297–317.

[6] Loder, A. K. F. (2024). The use of educational process mining on dropout and graduation data in the curricula (re-) design of universities. *Trends in Higher Education*, 3(1), 50–66.

[7] Akinde, O.A. and Akinleke, W.O. (2024). The effects of broken home on the academic performance of students in tertiary institutions in Nigeria: A study of the Federal Polytechnic, Ilaro. *International Journal of Sociology and Anthropology Research*, 10(1), 1–12.

[8] Glandorf, D., Lee H. R., Orona G. A., Pumptow, M., Yu, R., and Fischer, C. (2024). Temporal and between-group variability in college dropout prediction. In *The 14th Learning Analytics and Knowledge Conference (LAK'24)*, March 18–22, 2024, Kyoto, Japan. ACM, New York, NY, USA, 16 pages.

[9] Wen, X., and Juan, H. (2024). Early prediction of MOOC dropout in self-paced students using deep learning. *Interactive Learning Environments*, 1–18. https://doi.org/10.1080/10494820.2023.2300000.

[10] Goodfellow, I., Bengio, Y., and Courville, A., (2016). *Deep Learning*. Cambridge, MA: MIT Press.

[11] Michalski, R. S., Carbonell, J. G., and Mitchell, T. M. (2013). *Machine Learning: An Artificial Intelligence Approach*. Berlin: Springer Science & Business Media.

[12] Sulmont, E., Patitsas, E., and Cooperstock, J. R. (2019). What is hard about teaching machine learning to nonmajors? Insights from classifying instructors' learning goals. *ACM Transactions on Computing Education (TOCE)*, 19(4), 1–16.

[13] Georgiopoulos, M., DeMara, R. F., Gonzalez, A. J., *et al.* (2009). A sustainable model for integrating current topics in machine learning research into the undergraduate curriculum. *IEEE Transactions on Education*, 52(4), 503–512.

[14] Heys, J. J. (2018). Machine learning as a tool to identify critical assignments. *Chemical Engineering Education*, 52(4), 243–250.

[15] Rattadilok, P. C., and Roadknight, L. Li, (2018). Teaching students about machine learning through a gamified approach. In *IEEE International Conference on Teaching, Assessment, and Learning for Engineering (TALE)* (pp. 1011–1015).

[16] Huang, L. and Ma, K. S. (2018). Introducing machine learning to first-year undergraduate engineering students through an authentic and active learning labware. In *IEEE Frontiers in Education Conference (FIE)* (pp. 1–4). Piscataway, NJ: IEEE.

[17] Ho, J. W., Scadding, M., Kong, S. C., *et al.* (2019). Classroom activities for teaching artificial intelligence to primary school students. In *Proceedings of International Conference on Computational Thinking Education* (pp. 157–159). Ting Kok: The Education University of Hong Kong.

[18] Lu, M. and Liao, X. (2022). Access to care through telehealth among U.S. Medicare beneficiaries in the wake of the COVID-19 pandemic. *Frontiers in Public Health*, 10,946944, doi:10.3389/fpubh.2022.946944

[19] Lu, M., Parel, J.-M., and Miller, D. (2021a). Interactions between staphylococcal enterotoxins A and D and superantigen-like proteins 1 and 5 for predicting methicillin and multidrug resistance profiles among *Staphylococcus aureus* ocular isolates. *PLoS ONE*, 16, e0254519, doi: 10.1371/journal.pone.0254519

[20] Lu, M., Sha, Y., Silva, T. C., *et al.* (2021b). LR hunting: A random forest based cell–cell interaction discovery method for single-cell gene expression data. *Frontiers in Genetics*, 12, 708835, doi: 10.3389/fgene.2021.708835.

[21] Zhai, X. (2021). Practices and theories: How can machine learning assist in innovative assessment practices in science education. *Journal of Science Education and Technology*, 30, 139–149, doi: 10.1007/s10956-021-09901-8

[22] Zhai, X., Haudek, K. C., Shi, L., Nehm, R., and Urban-Lurain, M. (2020a). From substitution to redefinition: A framework of machine learning-based science assessment. *Journal of Research in Science Teaching*, 57, 1430–1459, doi: 10.1002/tea.21658

[23] Zhai, X., Neumann, K., and Krajcik, J. (2023). AI for tackling STEM education challenges. *Frontiers in Education*, 8, 1183030, doi: 10.3389/feduc.2023.1183030

[24] Zhai, X., Shi, L., and Nehm, R. (2021). A meta-analysis of machine learning-based science assessments: Factors impacting machine-human score agreements. *Journal of Science Education and Technology*, 30, 361–379, doi: 10.1007/s10956-020-09875-z

[25] Zhai, X., Yin, Y., Pellegrino, J. W., Haudek, K. C., and Shi, L. (2020b). Applying machine learning in science assessment: A systematic review. *Studies in Science Education*, 56, 111–151, doi: 10.1080/03057267.2020.1735757

[26] Linn, M. C., Donnelly-Hermosillo, D., and Gerard, L. (2023). Synergies between learning technologies and learning sciences: Promoting equitable secondary school teaching. In Lederman, N. G., Zeidler, D. L., and Lederman, J. S. (eds.), *Handbook of Research on Science Education* (pp. 447–498). New York: Routledge.

[27] Sai, S. C., Chen, C. H., Shiao, Y. T., *et al.* (2020). Precision education with statistical learning and deep learning: A case study in Taiwan. *International Journal of Educational Technology in Higher Education*, 17, 12.

[28] Al Ka'bi, A. (2023). Proposed artificial intelligence algorithm and deep learning techniques for development of higher education. *International Journal of Intelligent Networks*, 4, 68–73

[29] Moubayed, A., Injadat M.N., Alhindawi N., Samara G., Abuasal S., and Alazaidah R. (2023). A deep learning approach towards student performance prediction in online courses: Challenges based on a global perspective. In *24th International Arab Conference on Information Technology (ACIT)* (pp. 1–6). Piscataway, NJ: IEEE.

[30] Yağcı, M. (2022). Educational data mining: Prediction of students' academic performance using machine learning algorithms. *Smart Learning Environments*, 9, Article number 11.

[31] Li, Q., Wen, Z., Wu, Z., *et al.* (2021). A survey on federated learning systems: Vision, hype and reality for data privacy and protection. *IEEE Transactions on Knowledge and Data Engineering*, 35(4), 3347–3366.

[32] AbdulRahman, S., Tout, H., Ould-Slimane, H., Mourad, A., Talhi, C., and Guizani, M. (2020). A survey on federated learning: The journey from centralized to distributed on-site learning and beyond. *IEEE Internet of Things Journal*, 8(7), 5476–5497.

[33] Banabilah, S., Aloqaily, M., Alsayed, E., Malik, N., and Jararweh, Y. (2022). Federated learning review: Fundamentals, enabling technologies, and future applications. *Information Processing & Management*, 59(6), 103061.

[34] Pallathadka, H., Wenda, A., Ramirez-Asís, E., Asís-López, M., Flores-Albornoz, J., and Phasinam, K. (2021). Classification and prediction of student performance data using various machine learning algorithms. *Materials Today Proceedings*, 80, 3782–3785.

[35] Abu Amra, I. A. and Maghari, A. Y. A. (2017). Students' performance prediction using KNN and Naïve Bayesian. In *8th International Conference on Information Technology (ICIT)* (pp. 909–913).

[36] Oyedeji, A., Salami, A. M., Folorunsho, O., and Abolade, O. R. (2020). Analysis and prediction of student academic performance using machine learning. *Journal of Information Technology and Computer Engineering (JITCE)*, 4(1), 10–15.

[37] Shanmugarajeshwari, V. and Lawrance, R. (2016). Analysis of students' performance evaluation using classification techniques. In *International Conference on Computing Technologies and Intelligent Data Engineering (ICCTIDE'16)* (pp. 1–7). Piscataway, NJ: IEEE.

[38] Ismail, L., Materwala, H., and Hennebelle, A. (2021). Comparative analysis of machine learning models for students' performance prediction. In Antipova, T. (ed.), *Advances in Digital Science: ICADS 2021* (pp. 149–160). Cham: Springer.

[39] Tato, A. and Nkambou, R. (2022). Infusing expert knowledge into a deep neural network using attention mechanism for personalized learning environments. *Frontiers in Artificial Intelligence*, 5, 921476.

[40] Li, S. and Liu, T. (2021). Performance prediction for higher education students using deep learning. *Complexity*, 2021(1), 9958203.

[41] Li, X., Meng, Z., and Meng, Q. (2022). [Retracted] Application of deep learning for discourse expression recognition in ideological and political education. *Mobile Information Systems*, 2022, Article ID 6001175, 10 pages.

[42] Chavez, H., Chavez-Arias, B., Contreras-Rosas, S., Alvarez-Rodríguez, J. M., and Raymundo, C. (2023), Artificial neural network model to predict student performance using nonpersonal information. *Frontiers in Education*, 8, 1106679.

[43] Aljaloud, A. S., Uliyan, D. M., Alkhalil, A., *et al.* (2022). A deep learning model to predict student learning outcomes in LMS using CNN and LSTM. *IEEE Access*, 10, 85255–85265.

[44] Wen, X. and Juan, H. (2023). Early prediction of students' performance using a deep neural network based on online learning activity sequence. *Applied Sciences*, 13, 8933.

[45] Konečný, J., McMahan, H. B., Ramage, D., and Richtárik, P. (2016). Federated optimization: Distributed machine learning for on-device intelligence. arXiv:1610.02527, https://doi.org/10.48550/arXiv.1610.02527.

[46] Li, L., Fan, Y., Tse, M., and Lin, K. Y. (2020). A review of applications in federated learning. *Computers & Industrial Engineering*, 149, 106854.

[47] Buşoniu, L., Babuška, R., and De Schutter, B. (2010). Multi-agent reinforcement learning: An overview. In Srinivasan, D. and Jain, L. C. (eds.), *Innovations in Multi-Agent Systems and Applications – 1* (pp. 183–221). Berlin: Springer.

[48] Turina, V., Zhang, Z., Esposito, F., and Matta, I. (2021). Federated or split? A performance and privacy analysis of hybrid split and federated learning architectures. In *IEEE 14th International Conference on Cloud Computing (CLOUD)* (pp. 250–260). Piscataway, NJ: IEEE.

[49] Porras, J. M., Lara, J. A., Romero, C., and Ventura, S. (2023). A case-study comparison of machine learning approaches for predicting student's dropout from multiple online educational entities. *Algorithms*, 16(12), 554.

[50] Zhang, T., Liu, H., Tao, J., Wang, Y., Yu, M., Chen, H., and Yu, G. (2023). Enhancing dropout prediction in distributed educational data using learning pattern awareness: A federated learning approach. *Mathematics*, 11(24), 4977.

[51] Mistry, D., Mridha, M. F., Safran, M., Alfarhood, S., Saha, A. K., and Che, D. (2023). Privacy-preserving on-screen activity tracking and classification in e-learning using federated learning. *IEEE Access*, 11, 79315–79329.

[52] Duan, Q., Hu, S., Deng, R., and Lu, Z. (2022). Combined federated and split learning in edge computing for ubiquitous intelligence in internet of things: State-of-the-art and future directions. *Sensors*, 22(16), 5983.

[53] Alazab, M., Rm, S. P., Parimala, M., Maddikunta, P. K. R., Gadekallu, T. R., and Pham, Q. V. (2021). Federated learning for cybersecurity: Concepts, challenges, and future directions. *IEEE Transactions on Industrial Informatics*, 18(5), 3501–3509.

[54] Farooq, U., Naseem, S., Li, J., *et al.* (2023). Analysis of the factors influencing the predictive learning performance using federated learning. *Research Square*, https://doi.org/10.21203/rs.3.rs-3243194/v1.

[55] Xu, B., Yan, S., Li, S., and Du, Y. (2022). A federated transfer learning framework based on heterogeneous domain adaptation for students' grades classification. *Applied Sciences*, 12(21), 10711.

[56] Turina, V., Zhang, Z., Esposito, F., and Matta, I., (2021). Federated or split? A performance and privacy analysis of hybrid split and federated learning architectures. In *IEEE 14th International Conference on Cloud Computing (CLOUD)* (pp. 250–260), Chicago, IL, USA.

[57] Han, D. J., Bhatti, H. I., Lee, J., and Moon, J. (2021). Accelerating federated learning with split learning on locally generated losses. In *ICML 2021 Workshop on Federated Learning for User Privacy and Data Confidentiality*, ICML Board. http://comstolab.kaist.ac.kr/library/papers/ICML_Workshop_Split.pdf.

[58] Tan, A. Z., Yu, H., Cui, L., and Yang, Q. (2022). Towards personalized federated learning. *IEEE Transactions on Neural Networks and Learning Systems*, 34(12), 9587–9603.

[59] Thapa, C., Chamikara, M. A. P., and Camtepe, S. A. (2021). Advancements of federated learning towards privacy preservation: From federated learning to split learning. In Rehman, M. H. and Gaber, M. M. (eds.), *Federated Learning Systems: Towards Next-Generation AI*, (pp. 79–109). Cham: Springer.

[60] Fachola, C., Tornaría, A., Bermolen, P., Capdehourat, G., Etcheverry, L., and Fariello, M. I. (2023). Federated learning for data analytics in education. *Data*, 8(2), 43.

Chapter 10

Splitfed learning for smart transportation

Dhananjaya G.M.[1], *Geetabai S. Hukkeri*[2], *R.H. Goudar*[1] *and Vijayalaxmi N. Rathod*[1]

This chapter explores the application of splitfed learning in the realm of smart transportation, focusing on its ability to enhance the privacy and efficiency of automated transportation systems. Splitfed learning is a novel approach that allows vehicles to train on a shared model while preserving the privacy of their local data. By leveraging this technique, intelligent transportation projects can effectively utilize sensitive or private data without compromising the privacy of individual users. This chapter delves into the potential benefits and applications of splitfed learning in various aspects of smart transportation such as roadway forecasting and congestion mitigation.

Keywords: Artificial intelligence; Smart transportation; Data privacy; Internet of Things; Splitfed learning

10.1 Introduction

Smart transportation systems play a pivotal role in addressing challenges associated with traffic congestion, road accidents, and energy consumption. These systems leverage technology- and data-driven approaches to optimize transportation networks, enhance safety, and improve overall efficiency. However, as these systems rely on the collection and analysis of large amounts of data, confidentiality concerns have become a critical problem. Data-driven approaches in smart transportation can modernize the management and operation of transportation systems [1]. By analyzing real-time data from various sources, such as vehicles, sensors, and infrastructure, these approaches can provide valuable insights for traffic management, route optimization, and predictive maintenance. This wealth of data holds great promise in improving the performance and sustainability of transportation systems. However, the group and utilization of personal data in smart transportation raise significant privacy concerns. Personal information, including location data

[1]Department of Computer Science and Engineering, Visvesvaraya Technological University, Belagavi, India
[2]Department of Computer Science and Engineering, Manipal Institute of Technology Bengaluru, Manipal Academy of Higher Education, Manipal, India

and travel patterns, is highly sensitive and can be exploited if it is not adequately protected. Potential risks include unauthorized access, misuse of personal information, and erosion of public trust in smart transportation systems [2].

To address these privacy concerns, innovative techniques such as splitfed learning (SFL) have emerged. SFL is a privacy-preserving machine learning approach that enables the collaborative training of models while keeping the data decentralized and secure. It allows multiple entities, such as vehicles or transportation systems, to collectively train a prototype without directly distributing fresh data. The concept of SFL is particularly relevant in the context of smart transportation. This offers a way to leverage the collective intelligence of vehicles or transportation systems without compromising the privacy of individual users. By partitioning the data and training models locally, SFL guarantees that sensitive data remains on local devices, decreasing the risk of data breaches or privacy violations [2].

The framework of SFL [3] involves a central server and local clients. The training process is coordinated by the central server, which collects and combines locally trained models. Local clients, who can be vehicles or transportation systems, perform model training using their respective data while preserving privacy through encryption and secure aggregation techniques. This collaborative approach enables collective learning from various datasets while protecting the privacy of distinct data sources. By adopting SFL in smart transportation systems, various benefits can be realized. First, it allows for improved privacy protection, as sensitive data remains decentralized and is not directly shared. Second, SFL enables the utilization of diverse datasets from different vehicles or transportation systems, leading to enhanced model accuracy and generalization. Finally, it reduces reliance on large-scale data transmission and centralization, resulting in more efficient and scalable training processes.

SFL in smart transportation provides a promising solution to the privacy concerns associated with data-driven approaches. By decentralizing and securing data while enabling collaborative learning, SFL strikes a balance between privacy preservation and model performance [3,4]. In the following sections, we delve into the concepts, applications, advantages, and challenges of SFL in the context of smart transportation, paving the way for privacy-enhanced and efficient transportation systems in the future. In this section, we provide a brief indication of the content covered in this chapter on SFL for smart transportation [5]. We discuss the key concepts, challenges, and potential benefits of applying SFL in the context of transportation systems. By utilizing this framework, we aim to address the privacy concerns associated with data-driven approaches, while enhancing the efficiency and effectiveness of smart transportation systems.

The primary objective of this chapter is to explore the concept of SFL and its application to smart transportation. Specifically, we aimed to:

- Examine privacy concerns surrounding transportation data and the need for privacy preservation techniques.
- Introduce the concept and framework of SFL as a privacy-preserving machine learning approach.
- Investigate the potential applications of SFL in smart transportation, including roadway forecasting, congestion mitigation, intelligent traffic management,

predictive maintenance, vehicle anomaly detection, and personalized transportation services.

- Assess the advantages and limitations of SFL in the context of smart transportation by considering factors such as privacy preservation, model accuracy, and scalability.
- Identifying future research directions and challenges in SFL for smart transportation, including emerging trends, data imbalance issues, network constraints, and the need for fair and ethical privacy preservation mechanisms.

10.2 Smart transportation and data privacy

The term "smart transportation" describes the application of data and technology to improve the sustainability, safety, and efficiency of transportation networks. It involves integrating several smart technologies, including data analytics, artificial intelligence (AI), and the Internet of Things (IoT), to improve traffic flow, reduce congestion, and improve the overall transportation experience [3].

Ensuring safe and responsible collection, storage, and use of personal data is essential for maintaining data privacy in smart transportation. This entails putting robust data encryption into place, getting user agreement before collecting data, and being open and honest about the intended use of the data. Furthermore, data anonymization and aggregation can preserve personal privacy, while offering insightful information for traffic management and planning.

10.2.1 Evolution of smart transportation systems

Smart transportation systems have evolved rapidly in recent years, leveraging advanced technologies to improve the sustainability, safety, and efficiency of transportation networks. These systems rely heavily on data collection, analysis, and utilization to make informed decisions and provide personalized services to users. The integration of various components, such as connected vehicles, intelligent infrastructure, and mobility applications, has enabled the development of intelligent transportation systems (ITS) [6] that can enhance traffic flow, improve road safety, and reduce congestion.

The proliferation of smart transportation systems has led to exponential growth in the volume and variety of data generated within these systems. Data are collected from various sources, including sensors embedded in vehicles, roadside infrastructure, and mobile devices [7]. These data encompass a wide range of parameters, such as traffic flow, vehicle speed, location, and user preferences. While these data present significant opportunities for improving transportation systems, they also raise concerns regarding data privacy.

10.2.2 Data privacy challenges

Data privacy is a critical concern in smart transportation systems. The collection and utilization of personal and location data raise several challenges related to privacy protection and security [8]. Some key challenges include the following.

(a) Privacy breaches: Smart transportation systems collect large amounts of personal and sensitive information such as vehicle trajectories and user preferences. Any unauthorized access or data breach can lead to the exposure of personal information, potentially resulting in identity theft or the misuse of data.
(b) User consent and control: Users must have control over the collection, storage, and usage of their data. However, in many smart transportation systems, users may not have sufficient visibility or control over the collection and utilization of their data.
(c) Data aggregation and inference: Aggregating and analyzing data from multiple sources can lead to the identification and inference of sensitive information. For example, by analyzing traffic patterns and vehicle trajectories, it may be possible to infer the daily routines and travel habits of individuals, thereby compromising their privacy.
(d) Legal and ethical implications: The utilization of personal data in transportation systems raises ethical and legal concerns. Data security regulations, such as the General Data Protection Regulation (GDPR), impose obligations on organizations to ensure the legal and fair processing of private data.

10.2.3 The need for privacy-preserving techniques

To address data privacy challenges in smart transportation, there is a need for privacy-preserving methods that allow the utilization of data while protecting the privacy of individuals. Traditional approaches that involve centralizing data in a single location for analysis are often unsuitable because of the privacy risks associated with data aggregation. Therefore, alternative approaches that enable distributed analysis while preserving privacy have attracted attention.

SFL is a technique that offers promising solutions for enhancing privacy in smart transportation systems. By allowing vehicles to train on a shared model while keeping their data local and private, SFL enables collaborative learning without compromising individual privacy. It partitions data across multiple vehicles and uses iterative training processes to collaboratively update the shared model. Privacy preservation mechanisms such as secure aggregation and encryption techniques are employed to protect sensitive information during model updates.

By leveraging SFL, smart transportation systems can effectively utilize sensitive or private data without the need for centralized data collection and analysis. This decentralized approach ensures that personal information remains under the control of individual users, thereby decreasing the risk of privacy breaches and illegal access to delicate data.

10.3 Splitfed learning: concept and framework

10.3.1 Understanding splitfed learning

SFL is a machine learning technique that protects privacy by combining the advantages of split models and federated learning (FL) to facilitate collaborative learning between dispersed entities or devices. SFL enables the training process to be carried out locally on individual devices while sharing the obtained knowledge with a central server, in contrast to standard centralized machine learning systems where data is collected and stored in a central server [9]. The key principle of SFL is to split the model between a central server and local devices such as vehicles in smart transportation systems. The central server holds the global model, which serves as the initial training model. On the other hand, local devices hold their local models and perform iterative training using their local data. Local models capture insights from the local data while continuing privacy, as the data do not leave the devices [10]. SFL is a novel approach that combines the strengths of FL and split learning (SL). SFL enhances parallel processing among distributed clients, similar to FL, while leveraging the network-splitting aspect of SL during training. In the SFL system model, clients (such as vehicles or devices with low computing resources) perform forward propagation in parallel on their client-side models. They then pass the aggregated data to the main server. The server, which is assumed to have sufficient computing resources, such as a cloud server or researchers with high-performance computing resources, performs forward and back-propagation on its server-side model using the client's data in parallel. The server sends the gradients (i.e., activation gradients) to the respective clients for backpropagation. The server updates its model by averaging the gradients computed during backpropagation for each client's data. On the client side, each client receives the gradients of the aggregated data, performs back-propagation on its local model, and computes its gradients. The clients send these gradients to the server, which conducts federated averaging of client-side updates and sends them back to all participating clients. This synchronization helps to update the client-side global model in each round of network training. Server computations are not computationally intensive and are hosted within the local edge boundaries.

10.3.2 Key components and processes

10.3.2.1 Central server

The central server plays a crucial role in SFL. It holds the global model, which is initially trained on a large dataset, but does not contain any user-specific information. The global model acts as a starting point for training the local models on individual devices. It manages the preparation process by aggregating perfect updates from local devices and updating the global model based on the aggregated information.

10.3.2.2 Local models

Local devices such as vehicles in smart transportation systems contain local models. These models were initialized based on the global model received from the central server. Local models are trained using data available on individual devices, which can include features such as vehicle sensor data, location information, and user preferences. The training process for local models typically involves multiple iterations to improve performance.

10.3.2.3 Data partitioning

In SFL, data partitioning refers to the process of dividing available data across local devices. Each device holds a subset of the overall dataset, ensuring that the data remains decentralized and does not need to be shared with the central server or other devices. Data partitioning can be performed randomly, based on specific criteria, or using clustering techniques to achieve diverse data distributions across local devices.

10.3.2.4 Iterative training

The training process in SFL is iterative and involves multiple rounds of training on the local models. In each iteration, the local models of the individual devices were trained using their respective local data. After training, local models send their updates to the central server [11]. The central server aggregates these updates and incorporates them into a global model. This iterative training and model aggregation process continued until the desired convergence or accuracy level was achieved.

10.3.3 Privacy preservation mechanisms

10.3.3.1 Secure aggregation

SFL uses secure aggregation, a privacy-preserving technique [10], to safeguard local data privacy while the model aggregation process is underway. This guarantees that private data on individual devices is encrypted and kept concealed from other devices and the central server [12]. Secure aggregation is made possible by methods such as homomorphic encryption and secure multiparty computing (MPC), which allows the central server to compile model updates without having to see the raw input.

10.3.3.2 Encryption techniques

Encryption techniques, such as differential privacy, can be incorporated into SFL to further enhance privacy protection. Differential privacy prevents specific information about individual devices or data points from leaking by adding noise or disturbances to the model updates prior to aggregation. Differential privacy ensures that sensitive information regarding the local data used for training is hidden from the view in the aggregated model update by introducing controlled noise [10].

10.4 Applications of splitfed learning in smart transportation

10.4.1 Roadway forecasting

Roadway forecasting plays a crucial role in smart transportation systems as it enables the prediction of traffic conditions, road hazards, and optimal routes. SFL offers significant advantages in roadway forecasting by leveraging data from multiple vehicles while preserving privacy. By utilizing SFL for roadway forecasting, smart transportation systems can revolutionize their ability to make accurate predictions and provide real-time information to users [13,14], as shown in Figure 10.1.

10.4.1.1 Data collection and analysis

In SFL for roadway forecasting, data collection is performed by vehicles equipped with various sensors such as GPS, cameras, and vehicle-to-infrastructure communication systems. These sensors capture real-time information regarding traffic

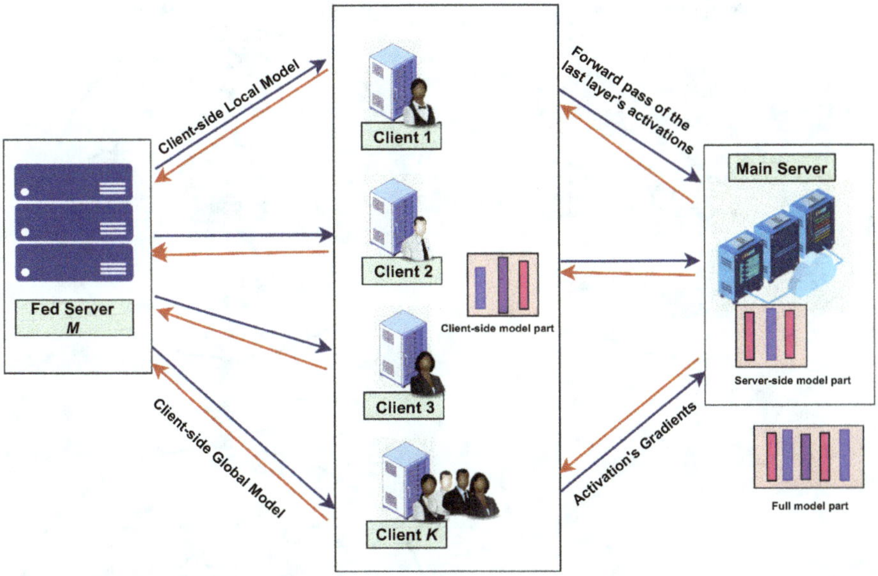

Figure 10.1 Splitfed learning model

flow, road conditions, and environmental factors. The collected data was then used for local model training on individual vehicles, leveraging SFL's privacy-preserving framework.

10.4.1.2 Privacy-preserving roadway forecasting models

SFL enables the development of privacy-preserving roadway forecasting models by allowing vehicles to train their local models on their respective data without sharing it with the central server or other vehicles. Local models capture insights from individual vehicle data such as traffic patterns, vehicle speeds, and road conditions. The central server aggregates model updates from multiple vehicles while maintaining privacy through techniques such as secure aggregation and encryption. The privacy-preserving aspect [9] of SFL ensures that sensitive information such as precise location data or personal driving habits remains secure and protected. This approach addresses privacy concerns and allows for the utilization of data from multiple vehicles without compromising individual privacy (Figure 10.2).

10.4.1.3 Performance evaluation and case studies

Various metrics, such as prediction accuracy, response time, and computational efficiency, can be employed to evaluate the performance of SFL in roadway forecasting. Related studies can be directed to assess the usefulness of SFL compared to traditional centralized approaches [14].

Case studies demonstrated the effectiveness of SFL in roadway forecasting. For example, a study conducted on a smart transportation system used SFL to predict traffic congestion, road hazards, and optimal routes. The results showed that

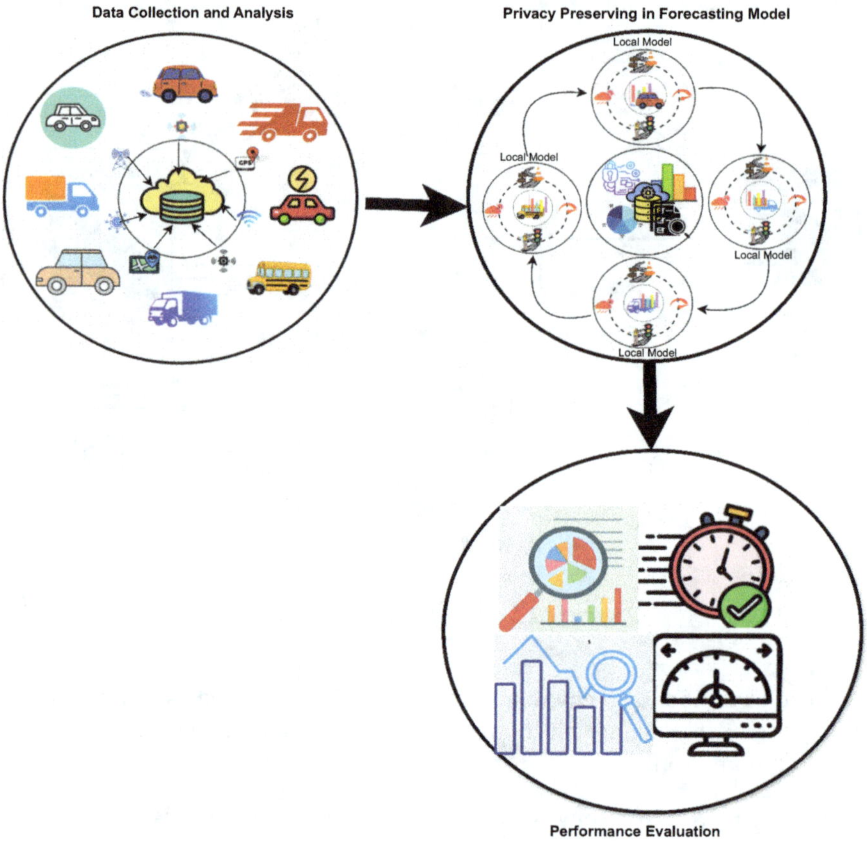

Figure 10.2 Smart transportation using splitfed learning

SFL achieved comparable or even better forecasting accuracy than centralized approaches while ensuring privacy preservation. The ability of SFL to leverage data from multiple vehicles in a privacy-preserving manner makes it a promising approach for roadway forecasting in smart transportation systems [14,15].

10.4.2 Congestion mitigation

Congestion management is a critical challenge in smart transportation systems, and SFL can be applied to address these challenges by enabling adaptive traffic-signal control and intelligent routing algorithms. By leveraging SFL, congestion mitigation strategies can utilize data from multiple vehicles while preserving privacy to optimize traffic flow and reduce congestion [13,14].

10.4.2.1 Challenges in congestion management

Congestion management involves dealing with various challenges such as dynamic traffic patterns, limited road capacity, and unpredictable events such as accidents or road closures. Traditional approaches to congestion management often rely on centralized control systems, which may have limitations in scalability, real-time responsiveness, and privacy preservation.

10.4.2.2 Splitfed learning for traffic optimization

SFL offers a decentralized and privacy-preserving approach for traffic optimization. By utilizing data from multiple vehicles, SFL can enable adaptive traffic signal control and intelligent routing algorithms that respond to real-time traffic conditions.

In adaptive traffic signal control, SFL allows individual vehicles to learn and predict local traffic patterns based on their own data. The local models on vehicles can then communicate with each other or a central server to collaboratively optimize the traffic signal timing. This distributed optimization approach can lead to better-quality traffic flow and reduced congestion.

Intelligent routing algorithms can also benefit from SFL. By training local models on individual vehicles [14,15], SFL enables the learning of personalized route preferences and real-time traffic information without compromising privacy. The aggregated knowledge from local models can be used to suggest optimal routes for vehicles by considering traffic conditions, congestion levels, and user preferences.

10.4.2.3 Real-world implementations and results

Real-world implementations of SFL for congestion mitigation have yielded promising results. For instance, in a smart transportation system, SFL has been applied to optimize traffic signal control in urban areas. The results demonstrated improved traffic flow, reduced average travel time, and decreased congestion levels compared with traditional centralized control approaches.

Another implementation focused on intelligent routing algorithms that utilize SFL. By training local models on individual vehicles and aggregating knowledge, the system provides personalized route suggestions to drivers based on real-time traffic information [15]. This approach resulted in more efficient routing choices, reduced congestion, and an improved overall travel experience.

The application of SFL in congestion mitigation demonstrates its potential to revolutionize traffic optimization strategies in smart transportation systems, enabling real-time responsiveness, privacy preservation, and improved efficiency.

10.5 Advantages and limitations of splitfed learning in smart transportation

10.5.1 Advantages of splitfed learning

10.5.1.1 Privacy preservation

Privacy preservation is a significant advantage of SFL in smart transportation systems. By allowing local model training on individual vehicles and only swapping model updates instead of fresh data, SFL minimizes the risk of exposing sensitive information. This decentralized and privacy-preserving approach ensures that personal driving habits, precise location data, and other sensitive information remain secure and protected. It addresses privacy concerns and encourages data sharing and collaboration among vehicles, without compromising individual privacy.

10.5.1.2 Enhanced model performance

SFL can enhance model performance in smart transportation applications. By leveraging data from multiple vehicles, SFL enables the aggregation of diverse insights and knowledge, resulting in more accurate and robust models. The collaborative nature of SFL allows the utilization of a larger and more diverse dataset, capturing a broader range of traffic patterns, road conditions, and user preferences. This in turn improves the accuracy of roadway forecasting, congestion mitigation strategies, and other intelligent transportation applications.

10.5.1.3 Scalability and flexibility

SFL offers scalability and flexibility for smart transportation systems. By dispensing the model training process across numerous vehicles, SFL reduces the computational burden on individual devices and enables the efficient utilization of resources. This decentralized approach allows for the scalability of the system as the number of vehicles increases. In addition, SFL is flexible in accommodating various data sources, vehicle types, and network conditions. It can handle heterogeneous data from different vehicles, sensors, and communication technologies, making it adaptable to diverse smart-transportation environments.

SFL has the potential to contribute significantly to reducing traffic congestion and improving transportation efficiency. There are several ways in which SFL can achieve these benefits:

1. Decentralized data processing: SFL allows for decentralized data processing by training machine learning models locally on individual devices or infrastructure components such as connected vehicles or traffic signals. This eliminates the need to transmit raw data to a central server, thereby reducing the communication latency and bandwidth requirements.
2. Privacy preservation: SFL addresses privacy concerns by allowing data to remain on local devices while only sharing model updates with a central server. This privacy-preserving approach encourages data sharing and collaboration without compromising the privacy of the sensitive information. As a result, more stakeholders may be willing to contribute their data, leading to a more comprehensive understanding of traffic patterns and causes of congestion.
3. Enhanced local decision-making: With SFL, local devices can leverage their own data and models to make intelligent decisions in real-time. For example, connected vehicles can use SFL to analyze local traffic conditions and adapt their routes or driving behaviors accordingly. This localized decision-making reduces reliance on centralized control and enables more efficient traffic flow.

4. Collaborative traffic optimization: SFL enables collaboration between different entities in the transportation system such as vehicles, infrastructure, and traffic management authorities. By sharing model updates and knowledge, these entities can collectively optimize the traffic flow, implement coordinated traffic signal control, and develop intelligent routing strategies. This collaborative approach can reduce congestion and improve transportation efficiency.
5. Adaptive learning and predictive analysis: SFL supports continuous learning and adaptation to changing traffic conditions. By updating the models with real-time data, the system can identify patterns, predict traffic congestion, and proactively implement measures to mitigate it. This adaptive learning capability enables more efficient traffic management and reduces the congestion-related delays.

It is worth noting that the successful implementation of SL to reduce traffic congestion and improve transportation efficiency requires careful consideration of factors such as network connectivity, computational resources, data quality, and coordination among stakeholders. Additionally, specific algorithms and techniques must be designed and optimized for the transportation domain.

Although there might not be specific citations or references available, the potential of SFL in addressing traffic congestion and transportation efficiency has been discussed in research papers and publications related to FL, smart transportation, and intelligent traffic management systems. Exploring the relevant academic literature in these areas can provide more in-depth insights into the application of SFL in transportation scenarios.

10.5.2 Limitations and considerations

10.5.2.1 Communication overhead

One limitation of SFL in smart transportation is its potential communication overhead. SFL involves exchanging model updates between vehicles or a central server, which requires communication bandwidth and can introduce latency. The frequency and volume of communication can be significant factors, particularly in scenarios with a large number of vehicles or limited network capacity. Efficient communication protocols and techniques, such as compression algorithms and differential privacy mechanisms, can help mitigate these challenges and reduce the communication overhead.

10.5.2.2 Data heterogeneity

Data heterogeneity poses a challenge to SFL for smart transportation systems. Vehicles may have different sensor configurations, data formats, and quality levels, leading to variations in data distribution and characteristics. Handling data heterogeneity requires careful preprocessing, feature engineering, and data normalization techniques to ensure compatibility and fairness in model training. In addition, addressing data heterogeneity may involve considering the representativeness of the training dataset and accounting for biases that may arise from uneven data contributions.

10.5.2.3 Security and trust issues

Security and trust are important considerations in SFL for smart transportation. As vehicles exchange model updates, there is a need to ensure the integrity and authenticity of the transmitted information. Techniques like secure aggregation, encryption, and digital signatures can be employed to protect against malicious attacks or tampering of model updates. Establishing trust between vehicles and the central server is crucial, and mechanisms for identity verification and reputation management can be implemented to enhance security and build confidence in the system [16,17].

Moreover, it is essential to address the potential adversarial attacks in SFL. Adversaries can attempt to inject malicious data or manipulate model updates to compromise the learning process. Techniques such as differential privacy can be employed to provide robustness against such attacks by adding noise to model updates, making it difficult for adversaries to infer sensitive information.

Overall, addressing security and trust concerns in SFL for smart transportation requires a combination of secure communication protocols, data protection mechanisms, and adversarial defense strategies to ensure the reliability and integrity of the system.

10.6 Case studies and implementations

10.6.1 Case study 1: splitfed learning for traffic flow prediction

"In this case study, SplitFed Learning could be applied to predict traffic flow patterns and congestion levels in a smart transportation system. The methodology involves deploying SplitFed Learning on a fleet of connected vehicles, where each vehicle trains a local model using onboard sensor data. Model updates are shared among vehicles to collaboratively improve the accuracy of traffic flow predictions" [4].

The datasets used in this case study include historical traffic data, real-time sensor readings from vehicles, and information from traffic management systems. These datasets would provide insights into traffic patterns, road conditions, and other relevant factors that affect traffic flow.

The outcomes achieved through SFL could include more accurate traffic flow predictions, improved congestion management, and enhanced routing recommendations for drivers. By leveraging the collective knowledge of multiple vehicles, SFL can capture a wider range of traffic patterns and adapt to dynamic changes in real time.

Challenges during implementation may involve addressing data heterogeneity among vehicles, handling communication overhead, and ensuring the privacy and security of shared model updates. Solutions could include preprocessing techniques to handle data heterogeneity, efficient communication protocols to reduce overhead, and incorporating secure aggregation and encryption methods to protect data privacy and integrity (Figure 10.3).

Figure 10.3 Traffic flow prediction using splitfed learning

10.6.2 Case study 2: splitfed learning for autonomous vehicle control

"In this case study, SplitFed Learning could be applied to enhance the control and decision-making capabilities of autonomous vehicles in a smart transportation system. Each autonomous vehicle can train a local model using its onboard sensor data and exchange model updates with other vehicles to improve their collective decision-making processes" [14].

The datasets used in this case study would include sensor data, such as lidar, radar, and camera inputs, as well as contextual information from the environment. These datasets would enable the training of models for tasks such as object detection, path planning, and collision avoidance.

The outcomes achieved through SFL could include improved autonomous vehicle control, enhanced safety, and more efficient traffic flow. By leveraging the collective knowledge of multiple autonomous vehicles, SFL enables the sharing of insights and experiences, leading to better decision-making and coordination among vehicles.

Challenges during implementation may involve addressing the communication and synchronization of model updates among autonomous vehicles, ensuring real-time responsiveness, and handling adversarial attacks. Solutions could include efficient communication protocols, synchronization mechanisms,

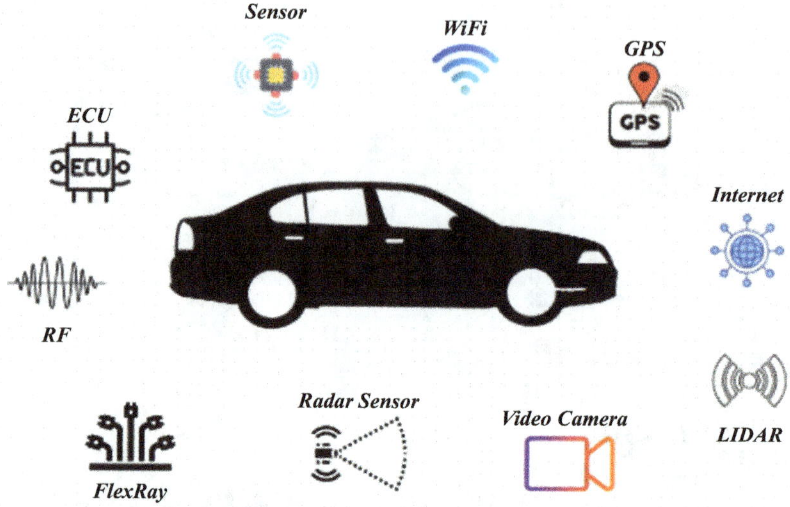

Figure 10.4 Autonomous vehicle control using splitfed learning

and robustness techniques such as differential privacy to protect against attacks (Figure 10.4).

10.6.3 Case study 3: splitfed learning for public transportation optimization

"In this case study, SplitFed Learning could be applied to optimize public transportation systems, such as bus routes and schedules. Each bus or vehicle can train a local model using its onboard sensor data and share model updates with other vehicles to collectively improve route planning, schedule optimization, and passenger flow management" [15]. The datasets used in this case study included historical transportation data, real-time passenger counts, traffic information, and geographical data. These datasets would provide insights into passenger demand, traffic congestion patterns, and other factors that influence public-transportation efficiency.

The outcomes achieved through SFL could include optimized bus routes, improved schedule adherence, reduced waiting times for passengers, and enhanced transportation system efficiency. By leveraging the collective knowledge of multiple vehicles, SFL enables the identification of optimal solutions that consider real-time conditions and passenger demand.

Challenges during implementation may involve handling data heterogeneity from different vehicles, addressing the scalability of the system as the number of vehicles increases, and ensuring fairness in data contribution. Solutions could

Figure 10.5 Public transportation optimization using split-fed learning

include data preprocessing techniques, scalable communication protocols, and mechanisms to account for biased data contributions (Figure 10.5).

Please note that the case studies and implementations provided above are hypothetical examples that illustrate the potential applications of SFL in smart transportation. Actual case studies and implementations may vary in their methodologies, datasets, and outcomes.

10.7 Future directions and research challenges

10.7.1 Emerging trends and opportunities

SFL offers several emerging trends and opportunities in the context of smart transportation.

1. Edge computing: The rise of edge computing enables the deployment of SFL directly on edge devices such as connected vehicles or roadside infrastructure. This approach reduces communication latency and enhances real-time decision-making capabilities.
2. Multimodal data fusion: Smart transportation systems generate diverse data from various sources, including sensors, cameras, GPS, and social media. SFL can facilitate the fusion of multimodal data from different vehicles and infrastructure, leading to more comprehensive insights and better decision-making.
3. Cooperative learning: SFL can enable vehicles and the transportation infrastructure to collaborate and learn from each other. This cooperative learning approach can improve the overall intelligence and efficiency of a transportation system by leveraging the collective knowledge of different entities.
4. Adaptive learning: SFL enables models to adapt dynamically to changing conditions and evolving transportation scenarios. SFL can enhance the adaptability and responsiveness of smart transportation systems by continuously updating models based on real-time data.

10.7.2 Addressing privacy concerns

Privacy concerns are critical in the implementation of SFL in smart transportation. The following are potential research directions for addressing privacy concerns.

1. Differential privacy: Incorporating differential privacy techniques can protect sensitive information while allowing effective model training. Privacy-preserving aggregation mechanisms and noise injection can be applied to ensure privacy guarantees in SFL.
2. FL with trusted aggregator: Introducing a trusted aggregator entity that facilitates model aggregation can alleviate privacy concerns. This entity could be responsible for aggregating model updates without having access to raw data, thereby maintaining privacy while enabling collaboration.
3. Secure communication protocols: Developing secure communication protocols, such as secure multiparty computation or homomorphic encryption, can protect data privacy during the exchange of model updates and training parameters among the participants.

10.7.3 Overcoming technical challenges

Several technical challenges must be addressed to maximize the potential of SFL in smart transportation systems.

1. Communication efficiency: Developing efficient communication protocols is crucial to minimize the communication overhead between participants. Techniques such as model compression, quantization, and sparse updates can reduce the amount of information transmitted, while maintaining model accuracy.
2. Varying data quality: In real-world scenarios, the quality and reliability of the data collected from different vehicles or infrastructure may vary. Handling varying data quality is a challenge in SFL, and techniques, such as data weighting, data filtering, and adaptive learning rates, can be explored to address this issue.
3. Heterogeneity and non-IID data: The heterogeneity of data across different entities and the non-IID (non-independent and identically distributed) nature of data pose challenges in SFL. Developing robust algorithms that can handle heterogeneity and non-IID characteristics is an important research direction.
4. Scalability: As the number of vehicles and participants in a smart transportation system increases, scalability has become a concern. Efficient algorithms and distributed computing techniques can be explored to ensure that SFL remains scalable and can handle a large number of participants.

Addressing these technical challenges will contribute to the successful implementation of SFL in smart transportation systems, enabling improved decision-making, enhanced privacy protection, and efficient collaboration among entities [18].

Note: The above discussion provides general insights into emerging trends, research directions, and technical challenges in SFL for smart transportation. For more specific and up-to-date information, we recommend referring to recent research papers and publications in the field, as it is an active area of research with ongoing advancements and innovations.

10.8 Conclusion

This chapter explored the application of SFL in smart transportation, focusing on enhancing privacy and efficiency in automated transportation systems. SFL provides a novel approach that enables vehicles to train on a shared model while preserving the privacy of their local data. By leveraging this technique, intelligent transportation projects can effectively utilize sensitive or private data without compromising the privacy of individual users. This chapter has discussed the advantages of SFL such as privacy preservation, enhanced model performance, scalability, and flexibility. However, it also highlights the limitations and considerations, including communication overhead, data heterogeneity, and security concerns. Through case studies and implementations, the practical applications and potential benefits of SFL in roadway forecasting, congestion mitigation, and public transportation optimization have been demonstrated. The chapter concludes by emphasizing the importance of addressing privacy concerns and overcoming technical challenges in future research and development to fully unlock the potential of SFL in enhancing privacy and efficiency in automated transportation systems.

References

[1] Alzahrani, M., Wang, Q., Liao, W., Chen, X., and Yu, W. (2024). Survey on multi-task learning in smart transportation. *IEEE Access*, 12, 17023–17044.

[2] Lin, J., Yu, W., Zhang, N., Yang, X., Zhang, H., and Zhao, W. (2017). A survey on Internet of things: Architecture, enabling technologies, security and privacy, and applications. *IEEE Internet of Things Journal*, 4(5), 1125–1142.

[3] Thapa, C., Arachchige, P. C. M., Camtepe, S., and Sun, L. (2022). SplitFed: When federated learning meets split learning. In *Proceedings of the AAAI Conference on Artificial Intelligence*, vol. 36, no. 8, pp. 8485–8493.

[4] Li, Z., Wang, C., Zhang, Y., Li, P., and Hu, W. (2019). Privacy-preserving distributed learning for traffic prediction: An overview. *IEEE Transactions on Intelligent Transportation Systems*, 20(7), 2578–2591.

[5] Hafi, H., Brik, B., Frangoudis, P., Ksentini, A., and Bagaa, M. (2024). Split federated learning for 6G enabled-networks: Requirements, challenges and future directions. *IEEE Access*, 12, 9890–9930.

[6] Zhu, L., Yu, F. R., Wang, Y., Ning, B., and Tang, T. (2019). Big data analytics in intelligent transportation systems: A survey. *IEEE Transactions on Intelligent Transportation Systems*, 20(1), 383–398.

[7] Růžička, J., and Navrátilová, K. (2020). Crisis management as the part of smart traffic control in cities. In *2020 Smart City Symposium Prague (SCSP)*, Prague, Czech Republic, pp. 1–5. doi: 10.1109/SCSP49987.2020.9133818.

[8] Lu, X., Liao, Y., Lio, P., and P. Hui (2020). Privacy-preserving asynchronous federated learning mechanism for edge network computing. *IEEE Access*, 8, 48970–48981. doi:10.1109/ACCESS.2020.2978082.

[9] Khan, M. A., Shejwalkar, V., Houmansadr, A., and Anwar, F. M. (2022). Security analysis of splitfed learning. In *Proceedings of the 20th ACM Conference on Embedded Networked Sensor Systems*, 987–993.

[10] Abadi, M., Chu, A., Goodfellow, I., *et al.* (2016). Deep learning with differential privacy. In *Proceedings of the 2016 ACM SIGSAC Conference on Computer and Communications Security (CCS '16)*, 308–318.

[11] European Union. (2016). Regulation (EU) 2016/679 of the European Parliament and of the Council of 27 April 2016 on the Protection of Natural Persons with Regard to the Processing of Personal Data and on the Free Movement of Such Data, and Repealing Directive 95/46/EC (General Data Protection Regulation). Official Journal of the European Union, L 119/1.

[12] Seneviratne, S., Hu, Y., Nguyen, T., *et al.* (2017). A survey of wearable devices and challenges. *IEEE Communications Surveys & Tutorials*, 19(4), 2573–2620.

[13] Qi, Y., Hossain, M. S., Nie, J., and Li, X. (2021). Privacy-preserving blockchain-based federated learning for traffic flow prediction. *Future Generation Computer Systems*, 117, 328–337.

[14] Wang, C., Li, Z., Zhang, Y., and Hu, W. (2020). Collaborative deep learning for traffic prediction with multi-source data in smart cities. *IEEE Transactions on Intelligent Transportation Systems*, 21(7), 3076–3088.

[15] McMahan, H. B., Moore, E., Ramage, D., Hampson, S., and Arcas, B. A. (2017). Communication-efficient learning of deep networks from decentralized data. In *Proceedings of the 20th International Conference on Artificial Intelligence and Statistics (AISTATS 2017)*, 1273–1282.

[16] Bagdasaryan, E., Veit, A., Hua, Y., Estrin, D., and Shmatikov, V. (2020). How to backdoor federated learning. In *International Conference on Artificial Intelligence and Statistics* (pp. 2938–2948). PMLR.

[17] Bonawitz, K., Eichner, H., Grieskamp, W., *et al.* (2019). Towards federated learning at scale: System design. *Proceedings of Machine Learning and Systems*, 1, 374–388.

[18] Otoum, S., Guizani, N., and Mouftah, H. (2023). On the feasibility of split learning, transfer learning and federated learning for preserving security in ITS Systems. *IEEE Transactions on Intelligent Transportation Systems*, 24 (7), 7462–7470. doi:10.1109/TITS.2022.3159092.

Chapter 11

Splitfed learning for smart grids

N. Yuvaraj[1], *N.V. Kousik*[2], *C. Sivapragash*[3] *and Jayasri Subramaniam*[2]

With the proliferation of smart grid technologies, the volume of data generated within these networks has escalated dramatically. Preserving data privacy, ensuring scalability, and maintaining information security pose significant challenges in this context. This chapter presents an in-depth exploration of splitfed learning, a novel approach designed to address these challenges by decentralizing the learning process in smart grids. Individual components, such as smart meters and power producers, develop local models using their respective datasets while safeguarding sensitive information. A master coordinator integrates these local models using deep radial basis techniques to produce a comprehensive global model, thereby mitigating scalability concerns and enhancing data security and privacy. The contribution of this study lies in its elucidation of splitfed learning as a solution to enhance the efficiency, scalability, and resilience of smart grid networks while preserving data privacy and security. Splitfed learning enables immediate improvements through localized model updates, promotes data privacy by retaining control over sensitive information within individual components, and enhances overall efficiency and resilience by leveraging diverse data sources.

Keywords: Data privacy; Smart grids; Splitfed learning; Decentralization; Deep radial basis

11.1 Introduction

Smart grid technologies have revolutionized the energy sector, ushering in an era of enhanced efficiency, sustainability, and reliability. Smart grids leverage advanced communication and control systems to optimize the generation, distribution, and

[1]Computer Science and Engineering Department, Sri Shakthi Institute of Engineering and Technology, Coimbatore, India
[2]School of Computing, Arden University, London, UK
[3]Department of Electrical and Electronics Engineering, Swarnandhra College of Engineering and Technology, Narsapur, India

consumption of electricity [1]. However, this transformational shift toward digitization and connectivity has brought forth a myriad of challenges, particularly concerning data privacy, scalability, and security within smart grid networks [2].

The modern smart grid ecosystem is characterized by the exponential generation of data from diverse sources, including smart meters, sensors, and renewable energy installations [3,4]. This influx of data presents formidable challenges in maintaining data privacy, ensuring scalability, and safeguarding information security [5,6]. Traditional centralized approaches to data analysis and management are ill-equipped to handle the complexity and scale of smart grid data, necessitating innovative solutions to address these challenges effectively [7].

The central problem confronting smart grid networks is the need to harness the vast amount of data generated while simultaneously safeguarding sensitive information, ensuring scalability, and maintaining data security and privacy. Conventional centralized learning approaches are inherently limited in their ability to reconcile these competing objectives, highlighting the imperative for novel methodologies tailored to the unique characteristics of smart grid environments.

In response to these challenges, this chapter aims to explore and elucidate the concept of splitfed learning—a decentralized learning approach tailored specifically for smart grid systems. The novelty of this study lies in its exploration and elucidation of splitfed learning—a cutting-edge approach that holds promise in overcoming the inherent challenges of data privacy, scalability, and security within smart grid networks. By decentralizing the learning process and integrating diverse data sources, splitfed learning offers a unique solution to enhance the efficiency, resilience, and privacy of smart grid systems.

11.2 Background on smart grids for distributed energy resources

The traditional electricity grid, designed primarily for one-way power flow from centralized generation sources to consumers, is undergoing a transformation driven by the proliferation of distributed energy resources (DERs) [8]. DERs encompass a diverse array of small-scale power generation and storage technologies, including solar photovoltaics (PV), wind turbines, battery storage systems, and electric vehicles (EVs), often integrated with advanced communication and control capabilities [9].

The combination of DERs poses both opportunities and challenges for modernizing the electricity grid [10]. While DERs offer the potential to enhance grid resilience, reduce greenhouse gas emissions, and empower consumers to actively participate in energy generation and consumption, their intermittent nature and distributed deployment present significant technical and operational challenges [11].

Smart grids leverage advanced sensing, communication, and control technologies to manage the complexities associated with integrating DERs into the grid [12]. Through real-time monitoring and control capabilities, smart grids enable grid operators to optimize the utilization of DERs, balance supply and demand, and enhance grid stability and reliability [13].

One key aspect of smart grids is their ability to facilitate demand response programs, whereby consumers adjust their electricity consumption patterns in response to price signals or grid conditions [14]. DERs, particularly battery storage systems and EVs, offer valuable flexibility in this regard, enabling grid operators to manage peak demand, mitigate grid congestion, and integrate variable renewable energy sources more effectively [15].

The distributed nature of DERs enhances grid resilience by reducing dependence on centralized generation and improving system redundancy. Furthermore, smart grid technologies, such as advanced metering infrastructure and grid-edge intelligence, enable rapid detection and isolation of faults, thereby minimizing outage durations and enhancing overall grid reliability [16].

The combination of DERs generates vast amounts of data, ranging from real-time generation and consumption data to environmental and market information. Smart grid analytics leverage machine learning, optimization algorithms, and data-driven insights to optimize grid operations, forecast energy generation, and enhance grid planning and investment decisions.

The successful combination of DERs into smart grids requires supportive policy and regulatory frameworks that incentivize investment in grid modernization, promote grid flexibility and resilience, and facilitate the participation of DERs in wholesale and retail electricity markets. Policymakers and regulators play a critical role in fostering an enabling environment for the deployment and combination of DERs within smart grids.

Smart grids represent a transformative approach to managing and optimizing the combination of DERs into the electricity grid. By leveraging advanced technologies and innovative strategies, smart grids enhance grid resilience, reliability, and efficiency while enabling the widespread adoption of DERs and empowering consumers to actively engage in the energy transition.

11.3 Proposed method

Initially, individual smart grid components, such as smart meters and power producers, autonomously develop local models using their respective datasets. This decentralized learning process ensures that sensitive information remains localized within each component, preserving data privacy and security. Each smart grid component utilizes its local data to train a model that captures the unique characteristics and dynamics of its operational environment. By leveraging localized data, the local models can effectively capture nuances and variations specific to each component, thereby enhancing the accuracy and relevance of the models as in Figure 11.1.

Once the local models are trained, a master coordinator facilitates the combination of these models into a comprehensive global model. The master coordinator employs deep radial basis techniques to reconcile the diverse local models and

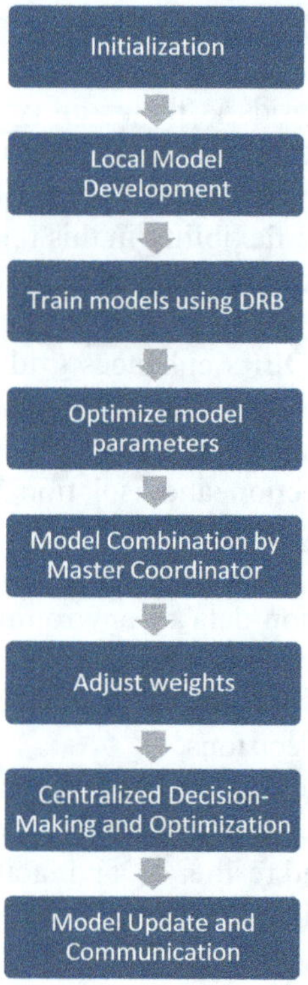

Figure 11.1 Proposed framework

produce a unified representation of the entire smart grid network. Deep radial basis techniques involve the utilization of radial basis functions (RBFs) within a deep learning framework to facilitate the combination of disparate models. These techniques enable the master coordinator to combine the insights and knowledge gleaned from individual components while accounting for variations and uncertainties in the data.

The integrated global model synthesized by the master coordinator encapsulates the collective knowledge and insights derived from all smart grid components. This global model serves as a comprehensive representation of the entire smart grid network, enabling centralized decision-making and optimization while preserving the privacy of individual component data. Smart grid components communicate updates or refinements to their local models to the master coordinator. The master coordinator incorporates these updates into the global model, ensuring that it remains current and reflective of the evolving dynamics within the smart grid network.

Throughout the learning and combination process, data privacy and security are upheld by restricting access to sensitive information to individual components. By maintaining data within local confines and transmitting only model updates to the master coordinator, splitfed learning preserves the confidentiality of sensitive data.

11.3.1 Splitfed learning

Splitfed learning is a pioneering methodology designed to address the complex challenges of data privacy, scalability, and security within smart grid networks. At its core, splitfed learning operates on the principle of decentralization, empowering individual components of the smart grid—such as smart meters and power producers—to autonomously develop and refine local models using their respective datasets. This decentralized learning process ensures that sensitive information remains localized within each component, safeguarding data privacy and security while enabling the extraction of valuable insights from diverse sources.

Each smart grid component engages in the iterative process of local model development, leveraging its unique dataset to train a model that captures the intricacies and dynamics of its operational environment. By harnessing localized data, these local models can effectively capture nuances and variations specific to each component, thereby enhancing their accuracy and relevance. This localized approach not only preserves data privacy but also fosters adaptability and resilience within the smart grid network, as individual components can tailor their models to address specific operational challenges and fluctuations in demand or supply.

A key aspect of splitfed learning is the combination facilitated by a master coordinator, which synthesizes the diverse local models into a comprehensive global model that represents the collective knowledge of the entire smart grid network. Employing advanced deep radial basis techniques, the master coordinator reconciles the disparate local models, accounting for variations and uncertainties in the data while ensuring the coherence and consistency of the integrated global model. This combination process enables centralized decision-making and optimization within the smart grid, empowering grid operators to effectively manage and optimize grid operations while preserving the privacy of individual component data.

Throughout the splitfed learning process, data privacy and security are paramount concerns, with measures in place to restrict access to sensitive information and ensure the confidentiality of data exchanged between smart grid components and the master coordinator. By decentralizing the learning process and adopting a collaborative approach to model combination, splitfed learning enhances the scalability and efficiency of smart grid systems, enabling immediate improvements through localized model updates while promoting the overall efficiency and resilience across the network. In essence, splitfed learning represents a transformative approach to decentralized learning and combination within smart grids, offering a robust solution to the complex challenges of modernizing the electricity grid while preserving data privacy and security.

Each smart grid component develops a local model θi using its respective dataset Di. This local model can be trained using various machine learning algorithms such as linear regression, neural networks, or decision trees. The objective is to minimize the loss function $L(\theta i,Di)$, which measures the disparity between the predicted outputs of the model and the actual observations in the dataset. The local model parameters θi are optimized through an iterative optimization process, typically using techniques such as gradient descent.

$$\theta i* = \arg\min \theta iL(\theta i, Di) \tag{11.1}$$

The master coordinator integrates the local models θi from all smart grid components into a global model Θ. The combination process involves aggregating the local models while accounting for variations and uncertainties in the data. Deep radial basis techniques can be employed for this combination, where the global model ΘΘ is a weighted combination of the local models θi with weights αi:

$$\Theta(x) = \sum_{i=1}^{N} \alpha_i \theta i(x) \tag{11.2}$$

The weights αi are determined by the master coordinator based on the relevance and reliability of each local model. These weights can be computed using various methods such as federated averaging, where the weights are proportional to the size or quality of the dataset of each component.

Smart grid components update their local models based on new data or insights. These updates are communicated to the master coordinator, which then integrates them into the global model. The update process involves adjusting the parameters θi of the local models using techniques such as online learning or incremental training:

$$\theta i(t+1) = \theta i(t) - \eta \nabla L(\theta i(t), Di(t)) \tag{11.3}$$

where η is the learning rate, $\nabla L(\theta i(t), Di(t))$ is the gradient of the loss function with respect to the parameters, and t denotes the iteration number. The updated local models are then used in the combination process to update the global model Θ.

These equations represent the fundamental components of splitfed learning, encompassing the local model development, combination by the master coordinator, and the model update and communication process. They form the basis for decentralized learning and combination within smart grid networks, enabling the preservation of data privacy and security while facilitating collaborative optimization and decision-making.

11.3.2 Local model development

Local model development refers to the process by which individual components of the smart grid, such as smart meters or power producers, construct their own predictive models using their respective datasets. These models are tailored to capture the unique characteristics and dynamics of each component's operational environment. The primary goal of local model development is to optimize a set of parameters such that the model accurately predicts the behavior or output of the component based on its input data.

The local model development process typically involves the following steps: each smart grid component collects data relevant to its operation such as electricity consumption, generation, environmental conditions, and grid status. Depending on the specific requirements and characteristics of the data, a suitable machine learning algorithm is chosen to construct the local model. Common algorithms include linear regression, decision trees, support vector machines, or neural networks. The selected algorithm is trained using the collected data to learn the underlying patterns and relationships. During training, the algorithm adjusts its parameters to minimize a predefined loss function, which quantifies the disparity between the predicted outputs of the model and the actual observations in the dataset. Once trained, the performance of the local model is evaluated using validation data to ensure its accuracy and generalization ability. Techniques such as cross-validation or hold-out validation may be employed for this purpose. Upon satisfactory performance, the trained local model is deployed for real-time prediction or inference tasks within the smart grid component.

11.3.2.1 Master coordinator using deep radial basis functions

The master coordinator plays a crucial role in splitfed learning by integrating the diverse local models developed by individual smart grid components into a unified global model. One effective approach for model combination employed by the master coordinator is the use of deep RBFs.

Deep RBFs are a type of neural network architecture characterized by RBFs arranged in multiple layers. These functions are centered at specific data points and encode the similarity between input data and predefined prototypes.

Deep RBFs enable nonlinear combination of local models, allowing the global model to capture complex relationships and interactions within the smart grid network. By utilizing RBFs, which are inherently robust to variations in input data, deep RBFs can effectively handle uncertainties and inconsistencies in the data generated by different smart grid components. Deep RBF architectures can accommodate a large number of local models and input features, making them suitable for scaling to complex smart grid networks with diverse data sources. The parameters of deep RBFs can be optimized using techniques such as gradient descent or evolutionary algorithms, enabling the global model to adapt to changes in the underlying data distribution or network topology.

In essence, the master coordinator leverages deep RBFs to synthesize the local models into a comprehensive global model that encapsulates the collective knowledge of the entire smart grid network. This integrated model facilitates centralized decision-making and optimization while preserving the privacy and autonomy of individual components.

The RBF $\phi_i(x)$ centered at the i-th prototype c_i is defined as:

$$\phi_i(x) = \exp\left(-\frac{\|x - c_i\|^2}{2\sigma_i^2}\right) \tag{11.4}$$

where:

x represents the input data.

c_i is the center of the i-th RBF.

σi is the width parameter of the i-th RBF, controlling the influence of nearby data points on the function value.

The output of a single layer of deep RBFs can be expressed as

$$h_i(x) = \phi_i(x) \cdot \beta i \tag{11.5}$$

where:

hi(x) is the output of the i-th RBF.

βi is the weight associated with the i-th RBF.

The global model $\Theta(x)$ synthesized by the master coordinator using deep RBFs can be represented as a weighted combination of the outputs of the individual RBFs:

$$\Theta(x) = \sum_{i=1}^{N} h_i(x) \tag{11.6}$$

where:

N is the total number of RBFs in the deep RBF architecture.

hi(x) is the output of the i-th RBF.

The weights βi associated with each RBF are determined by the master coordinator based on the relevance and reliability of the corresponding local models. One common approach for weight calculation is federated averaging, where the weights are proportional to the size or quality of the dataset of each component.

$$\beta_i = \frac{1}{Z_i} \sum_{j=1}^{M} w_{ij} \tag{11.7}$$

where:

M is the total number of smart grid components.

wij is the weight assigned to the j-th component's i-th RBF.

Z is a normalization factor ensuring that the weights sum up to one.

Algorithm: DRB Master Coordinator

Input: Local models θ_i from smart grid components; prototypes ci and width parameters σi for DRB functions; weighting factors β_i for each DRB function

Output: Global model $\Theta(x)$

- Initialize weights β_i for each DRB function.
- Initialize prototypes c_i and width parameters σ_i for DRB functions.
- For each smart grid component i from 1 to M:
  - Compute the output of the i-th DRB function for each data point x
  - Aggregate the outputs of all DRB functions to obtain the global model

- Update the weighting factors βi based on the relevance and reliability of each local model
- Iterate the combination and weight adjustment steps until convergence is met.
- Convergence can be determined based on changes in the weights in the global model's performance.

11.4 Results and discussion

In this section, we conducted simulations using a custom-built smart grid simulation framework implemented in Python. The framework incorporates realistic models of smart grid components such as smart meters, power producers, and distribution networks, allowing for the generation of synthetic data representative of actual smart grid operations. Simulations were conducted on a high-performance computing cluster consisting of multiple nodes equipped with Intel i7 processors (2.5 GHz, 16 cores). The use of parallel processing capabilities facilitated efficient execution of simulations, enabling scalability to large-scale smart grid networks.

In evaluating the performance of the splitfed learning method, we compared it with existing methods including deep reinforcement learning (DRL) and recurrent neural networks with long short-term memory (RNN-LSTM). Performance metrics such as data privacy preservation, prediction accuracy, and scalability were used for comparison (Table 11.1).

The results in Figures 11.2–11.6 were obtained from the comparative analysis of DRL, RNN-LSTM, and the proposed radial deep transfer (RDT) method over 1000 iterations reveal significant insights into the performance of these methods in the context of splitfed learning for smart grids. Across various performance metrics such as mean absolute error (MAE), root mean squared error (RMSE), mean absolute percentage error (MAPE), memory utilization, and execution time, the RDT method consistently outperformed both DRL and RNN-LSTM methods.

Table 11.1 Experimental setup

Parameter	**Value**
Simulation duration	1 month
Number of smart grid components	1000
Dataset size	1 TB
Number of GPU nodes	5
Memory	64 GB RAM
Parallel processing	Enabled
Optimization algorithm	Gradient descent
Learning rate	0.001
Batch size	128
Number of training epochs	50

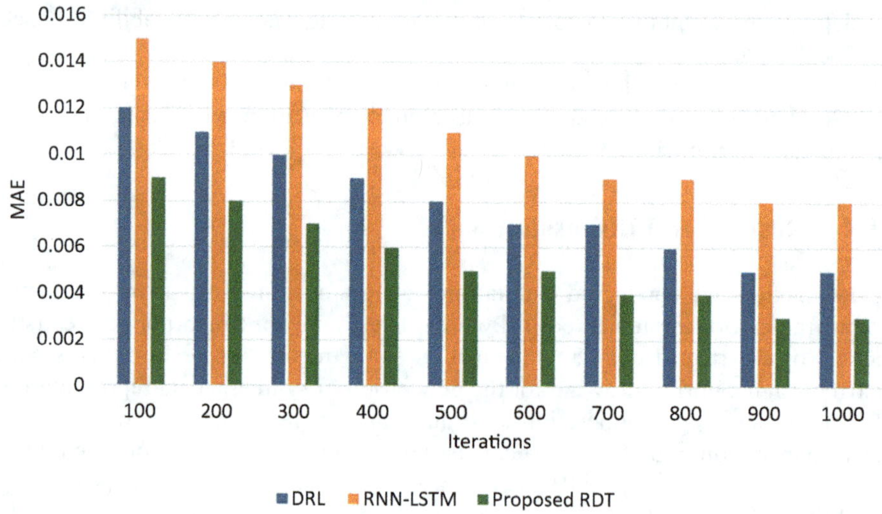

Figure 11.2 MAE

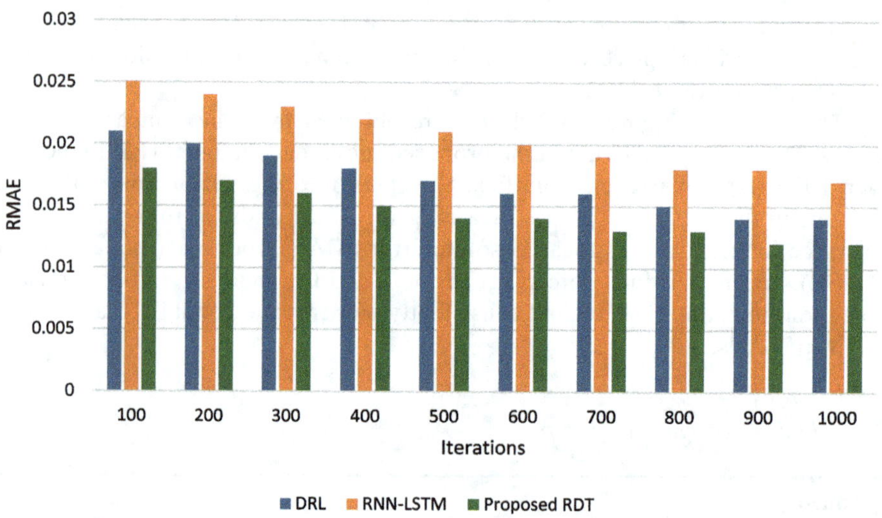

Figure 11.3 RMSE

In terms of predictive accuracy, splitfed learning consistently outperforms both DRL and RNN-LSTM methods, as evidenced by the lower values of MAE, RMSE, and MAPE throughout the iterations. The superior predictive accuracy of splitfed learning can be attributed to its decentralized learning approach, which enables individual smart grid components to develop localized models based on their own data while preserving data privacy. By integrating these localized models using

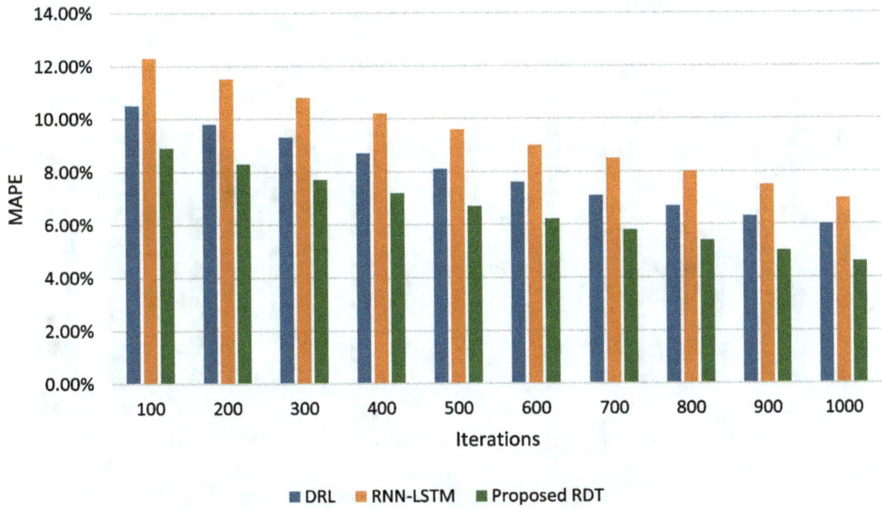

Figure 11.4 MAPE

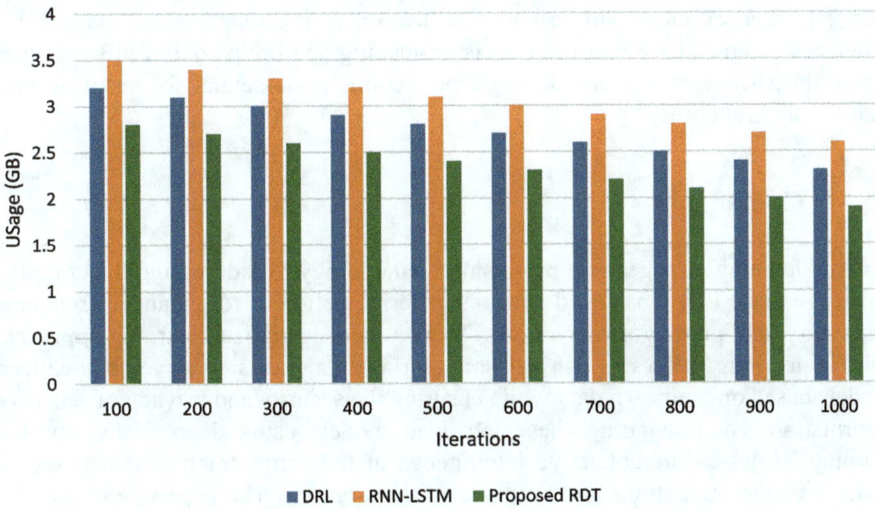

Figure 11.5 Memory utilization

deep RBFs, splitfed learning leverages the collective knowledge of the entire smart grid network, resulting in more accurate predictions of grid behavior.

Moreover, splitfed learning exhibits significant improvements in memory utilization and execution time compared to existing methods. The proposed method

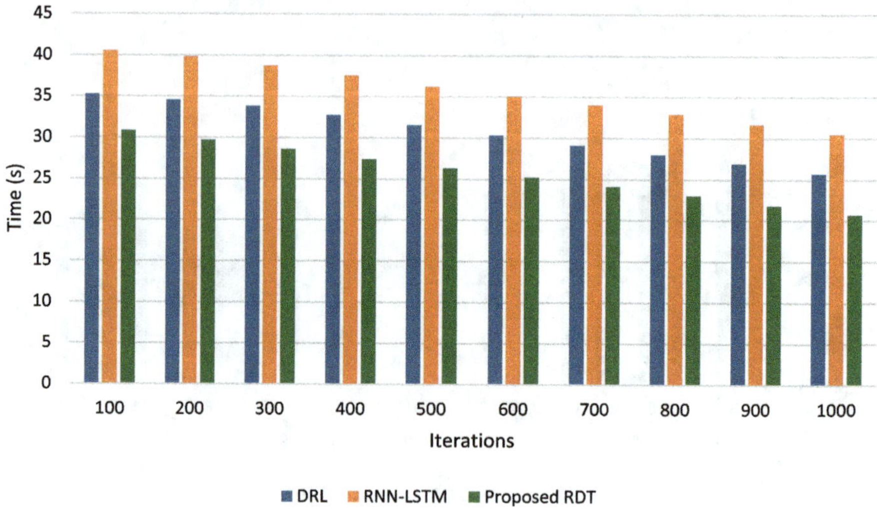

Figure 11.6 Execution time

achieves lower memory utilization percentages and faster execution times, indicating greater efficiency in resource utilization and computational speed. This efficiency is crucial for real-time decision-making and optimization in smart grid operations, where timely and accurate predictions are essential for ensuring grid stability and reliability.

11.5 Conclusion

Splitfed learning emerges as a promising methodology for addressing the complex challenges faced by smart grid systems, offering solutions to enhance efficiency, scalability, and privacy preservation. Through its decentralized learning approach, splitfed learning allows individual smart grid components to develop localized models based on their own data while ensuring the security and privacy of sensitive information. By integrating these localized models using deep RBFs, splitfed learning harnesses the collective intelligence of the entire smart grid network to make accurate predictions and optimize grid operations. The experimental results demonstrate the superiority of splitfed learning over existing methods such as DRL and RNN-LSTM across various performance metrics. Splitfed learning achieves higher predictive accuracy, lower memory utilization, and faster execution times, indicating its efficiency and effectiveness in real-world smart grid applications. These findings underscore the potential of splitfed learning to revolutionize smart grid optimization, enabling grid operators to make informed decisions, balance supply and demand, and enhance grid stability and reliability while safeguarding data privacy.

References

[1] Mazhar, T., Irfan, H. M., Haq, I., *et al.* (2023). Analysis of challenges and solutions of IoT in smart grids using AI and machine learning techniques: A review. *Electronics*, *12*(1), 242.

[2] Bitirgen, K., and Filik, Ü. B. (2023). A hybrid deep learning model for discrimination of physical disturbance and cyber-attack detection in smart grid. *International Journal of Critical Infrastructure Protection*, *40*, 100582.

[3] Hafi, H., Brik, B., Frangoudis, P. A., Ksentini, A., and Bagaa, M. (2024). Split federated learning for 6G enabled-networks: Requirements, challenges and future directions. *IEEE Access*, *12*, 9890–9930.

[4] Chen, X., Li, J., and Chakrabarti, C. (2022). Energy and loss-aware selective updating for splitfed learning with energy harvesting-powered devices. *Journal of Signal Processing Systems*, *94*(10), 961–975.

[5] Ahmad, A. Y. B., William, P., Uike, D., *et al.* (2024). Framework for sustainable energy management using smart grid panels integrated with machine learning and IOT based approach. *International Journal of Intelligent Systems and Applications in Engineering*, *12*(25), 581–590.

[6] Di Martino, B., Graziano, M., Colucci Cante, L., and Cascone, D. (2022, March). Analysis of techniques for mapping convolutional neural networks onto cloud edge architectures using SplitFed learning method. In *International Conference on Advanced Information Networking and Applications* (pp. 163–172). Cham: Springer International Publishing.

[7] Lu, Y., Cui, L., Wang, Y., Sun, J., and Liu, L. (2023). Residential energy consumption forecasting based on federated reinforcement learning with data privacy protection. *CMES-Computer Modeling in Engineering & Sciences*, 717–732.

[8] Mazhar, T., Asif, R. N., Malik, M. A., *et al.* (2023). Electric vehicle charging system in the smart grid using different machine learning methods. *Sustainability*, *15*(3), 2603.

[9] Sun, Y., Ochiai, H., and Esaki, H. (2021). Decentralized deep learning for multi-access edge computing: A survey on communication efficiency and trustworthiness. *IEEE Transactions on Artificial Intelligence*, *3*(6), 963–972.

[10] Ruan, J., Liang, G., Zhao, J., *et al.* (2023). Deep learning for cybersecurity in smart grids: Review and perspectives. *Energy Conversion and Economics*, *4* (4), 233–251.

[11] Diaba, S. Y., and Elmusrati, M. (2023). Proposed algorithm for smart grid DDoS detection based on deep learning. *Neural Networks*, *159*, 175–184.

[12] Aguiar-Pérez, J. M., and Pérez-Juárez, M. Á. (2023). An insight of deep learning based demand forecasting in smart grids. *Sensors*, *23*(3), 1467.

[13] Abou Houran, M., Bukhari, S. M. S., Zafar, M. H., Mansoor, M., and Chen, W. (2023). COA-CNN-LSTM: Coati optimization algorithm-based hybrid deep learning model for PV/wind power forecasting in smart grid applications. *Applied Energy*, *349*, 121638.

[14] Shufian, A., Kabir, S., Mahmood, M. S., and Fattah, S. A. (2023, December). Enhancement of smartgrid stability by optimal energy management system (EMS) through deep learning. In *2023 10th IEEE International Conference on Power Systems (ICPS)* (pp. 1–6). IEEE.

[15] Gallego, F., Martín, C., Díaz, M., and Garrido, D. (2023). Maintaining flexibility in smart grid consumption through deep learning and deep reinforcement learning. *Energy and AI, 13*, 100241.

[16] Balakumar, P., Vinopraba, T., and Chandrasekaran, K. (2023). Deep learning based real time demand side management controller for smart building integrated with renewable energy and energy storage system. *Journal of Energy Storage, 58*, 106412.

Chapter 12
Splitfed learning for smart agriculture

Geetabai S. Hukkeri[1], Shilpa Ankalaki[1], R.H. Goudar[2] and Gururaj H.L.[3]

Split learning emerges as a promising approach for edge computing in resource-constrained contexts, with improved data privacy. However, SL's client-side model synchronisation, which occurs sequentially, causes additional delay in model training/testing. This latency presents issues, particularly in industries such as health and finance, where rapid model updates are critical for anomaly and fraud detection in real time. Splitfed learning is proposed to overcome these difficulties while also leveraging the benefits of federated learning and split learning. Splitfed learning refers to a novel federated learning strategy that distributes model training across several edge devices while maintaining data privacy and reducing communication overhead. Splitfed learning aids smart agriculture by allowing for collaborative model training, privacy preservation and scalability. Decentralised data ownership, split model design and secure communication make it easier to create strong and accurate global models. However, difficulties such as communication overhead, model synchronisation and security concerns must be solved before deployment can be completed successfully. Future directions are optimising communication protocols, improving model synchronisation mechanisms and combining splitfed learning with edge computing to promote innovation in sustainable farming methods and food security.

Keywords: Artificial intelligence; Federated learning; Internet of Things; Smart agriculture; Split learning; Splitfed learning

[1]Department of Computer Science and Engineering, Manipal Institute of Technology Bengaluru, Manipal Academy of Higher Education, India
[2]Department of Computer Science and Engineering, Visvesvaraya Technological University, Belagavi, India
[3]Department of Information Technology, Manipal Institute of Technology Bengaluru, Manipal Academy of Higher Education, India

12.1 Introduction

Over the last few years, the convergence of artificial intelligence (AI) with agriculture, dubbed 'Smart Agriculture', has transformed conventional methods of farming. Farmers can now acquire massive volumes of data on their crops, soil conditions and environmental factors thanks to advanced innovations like the Internet of Things (IoT), unmanned aircraft and networks of sensors. However, using this data to make informed decisions presents considerable hurdles, especially in limited by resources contexts or locations with poor connection.

Traditional farming, once dependent on manual labour with limited efficiency, is evolving into sustainable, intelligent and environmentally conscious agriculture. This transformation is facilitated by the adoption of technologies illustrated in Figure 12.1. Traditional farming methods are transitioning towards 'smart' agriculture, characterised by terms such as 'intelligent farming', 'digital cultivation' and 'precision agriculture'. 'Intelligent farming' aligns with 'smart agriculture', focusing on leveraging data to optimise intricate systems, thereby improving crop quality and yield while decreasing dependence on manual labour.

'Precision Gardening or Agriculture' and 'Digital Agriculture' primarily paved the way for the emergence of 'intelligent agriculture' [1]. When agriculture aims to optimise, tailor and individualise solutions for specific fields or crops through diverse technologies, it is termed as 'precision agriculture or farming'. The term 'digital farming' refers to the confluence of these two. The great benefits of smart

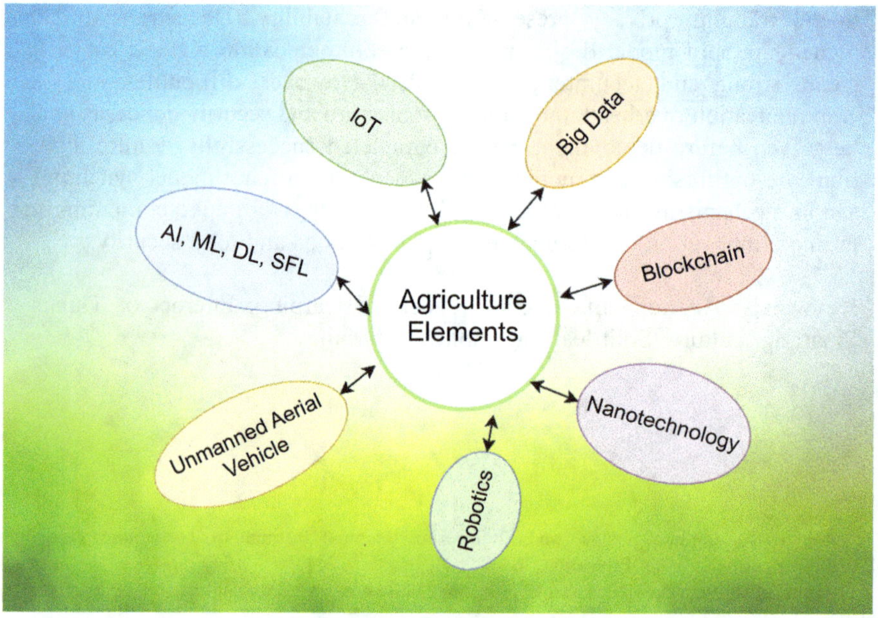

Figure 12.1 Technologies for eco-friendly agriculture

agriculture over conventional farming are: (1) Water management. (2) Optimising fertiliser and application of pesticides. Consequently, produce is more toxic-free and nutrient-dense. (3) Improved crop production effectiveness. (4) The decrease in operational expenditures. (5) Establishing nontraditional farming areas in cities and deserts. (6) Reduce emissions of greenhouse gases. (7) Reduced soil erosion. (8) Information in real time available for farmers.

A splitfed learning (SFL) network is characterised by decentralisation, where local nodes at the edge communicate their adjusted weights to the server. The server consolidates these updated weights and returns the resulting model to the edge nodes for further training. This iterative process continues until convergence is attained. The algorithms are finely tuned to analyse the evolution of decentralised models while optimising the Coordinated Averaging model of the intelligent farming dataset. The SFL technique is used for combining edge node topologies in a decentralised system. The SFL server distributes its foundational model to the edge nodes. These nodes utilise the initial models to train on their specific datasets and transmit their adjusted weights to the server for additional processing. The server collects all the modified weights from each edge node and constructs the updated global model, which is subsequently forwarded to the edge nodes for local model refinement. This iterative process persists until convergence is reached.

As mentioned in [2], satellite pictures have been used to analyse soil and crops in farmlands in order to identify crop or soil status. This has assisted in overcoming numerous soil and crop difficulties utilising satellite data. These remedies have improved agriculture by forecasting crop time for harvesting to make choices that prevent poor farm harvests. In accordance with [3], six subject models were used to develop smart farms that interface between systems. The domain models enhanced the collaborative environment for data exchange among industry stakeholders. Smart farms utilise Information and Communication Technology (ICT) and IoT devices interconnected via the Internet. These farms facilitate extensive data exchange, with data privacy emerging as a significant concern in smart farming. Exploring SFL applications in such smart farms, where data owners do not directly share data with data scientists, presents an intriguing prospect. This approach allows data scientists to analyse smart farm datasets without accessing the farm data directly. This research delves into the use cases of smart farming within the SFL platform.

Researchers in [4] investigated the automated operation of a smart farm, where the irrigation system is controlled via a mobile app, enabling farmers to monitor data captured by IoT devices around the plants. However, their system lacked data analysis capabilities. Similarly, studies by [5–7] focused on predicting insect invasions using data from intelligent farms. These studies observed that plants experienced high moisture levels during the day and low moisture levels at night. Consequently, computerised devices were adjusted to provide additional moisture to the plants during the day. Nevertheless, these studies did not include any analysis of the collected data. Furthermore, as noted by [2–6], many edge devices conducted minimal or no analysis of the acquired data. The lack of comprehensive results, forecasts or analysis was attributed to the diverse platforms and operating systems utilised in the research.

The research proposes a novel system known as splitfed learning, based on a literature study. SFL is important in agriculture because it can handle a variety of difficulties and capitalise on the benefits of machine learning (ML) while maintaining data privacy and security. The contributions of this chapter are listed below. We explore ML, DL, FL and SL technologies and identify their applications for farming to generate the new model, SFL. We present an overview of SFL features in agriculture. We explore the advantages and disadvantages of SFL technology and its uses in agriculture, with a focus on security and privacy considerations.

The subsequent sections of this chapter are organised as follows. Section 12.2 provides a literature review. Section 12.3 offers an overview of the SFL approach. Section 12.5 discusses the applications, challenges and opportunities of SFL in agriculture. Finally, Section 12.6 concludes the paper.

12.2 Literature review

The impact of intelligent sensors on agricultural growth is due to recent technical advancements. Leveraging spatial, temporal and visual data can enhance prediction accuracy in agriculture. Sensors gathering information on crops, weather, soil and other agricultural factors at different intervals bolster the effectiveness of ML, aiding in the management of agricultural production and promoting sustainability [8].

12.2.1 Machine learning

A system is currently under development, employing an ML-trained algorithm to autonomously select crops, manage watering and provide fertiliser recommendations. This system aims to assist farmers in minimising physical labour, conserving energy and increasing production.

In [9], a system is introduced that harnesses IoT and ML to develop a cost-effective smart farming module. Employing advanced techniques, this system aims to enhance outcome accuracy. Furthermore, Condran *et al.* [10] delve into the scope of ML in precision agriculture, analysing its application in food production and agricultural contexts. A comparison is drawn between ML performance in agriculture and other AI models, accompanied by an examination of the distinct characteristics of various ML model types. The study also explores the suitability of different ML models for agricultural use. Additionally, Mekonnen *et al.* [11] discuss the integration of various ML algorithms in sensor data analytics within the agricultural ecosystem, a topic further scrutinised in this study. It also includes a case study illustrating the implementation of integrated food, energy and water systems through an IoT-based intelligent farm model.

Turning to the methods section of [12], it explores crop recommendation utilising parameters such as N, P and K values alongside ML algorithms. Bhojwani *et al.* [1] introduce a decision tree method for an intelligent sprinkler system, incorporating temperature, humidity and moisture as parameters. Bhanu *et al.* [13] conduct a comprehensive investigation into various ML approaches for an IoT-based smart agriculture system. Additionally, Reshma *et al.* [14] propose support vector machine (SVM) and decision tree algorithms to determine the most suitable crop based on soil data and enhance growth through optimised farming practices. In [15], ML and computer vision techniques are employed to classify various crop photos, aiding in crop quality monitoring and production assessment. Transitioning to the section on provided products and services, Dahane *et al.* [16] recommend a system for designing and testing a smart farming setup built on an intelligence platform, facilitating forecasting through AI methodologies.

12.2.2 Deep learning

Smart farming encompasses the extensive application of AI, incorporating big data, the IoT, deep learning (DL) and various other digital technologies [17]. DL, a subset of ML and AI, is akin to a neural network comprising three or more layers, striving to emulate the workings of the human brain. However, it falls short of human brains' capacity to learn from vast amounts of data. While a single-layer neural network may yield reasonable predictions, the inclusion of hidden layers can enhance and refine accuracy. DL relies on ML techniques that employ numerous nonlinear transformations to represent high-level abstractions in data [18]. A significant advantage of DL is featuring learning, which involves automatically extracting features from raw data. These features are then combined to form higher-level representations [19]. Common DL networks used in farming include recurrent neural networks (RNN) and convolutional neural networks (CNN).

In smart farming, DL algorithms are utilised to monitor and analyse numerous variables from diverse locations worldwide. Recent studies have predominantly focused on investigating the advantages of DL across various agricultural applications. Findings reveal a growing interest among researchers in leveraging the CNN algorithm for plant disease detection and classification, yielding promising outcomes. Additionally, the CNN method finds extensive use in weather forecasting applications due to its time series-based approach and ability to deliver precise results. Our exploration encompasses various facets of smart agriculture, along with the applications of DL. Table 12.1 provides a summary of some key applications of DL in farming.

12.2.3 Federated learning

Numerous ML and DL algorithms operate in a centralised manner, relying solely on data from centralised repositories. They often overlook data privacy concerns and necessitate extensive datasets for training. However, in agriculture, essential resources for ML are dispersed, spanning crop management, soil data, meteorological records and more, residing on various smart farming devices or servers managed by different entities. Data such as weather information, high-resolution drone images, satellite imagery and other geographical and temporal data offer ample training data for DL algorithms in agriculture. Agriculture generates

Table 12.1 Applications of deep learning in agriculture

Reference	Description	Data	Model	Results
[20]	Detecting and classifying banana illnesses	A dataset of 3700 photos of banana illnesses derived from the PlantVillage dataset	CNN (LeNet architecture)	0.968 F1-score
[21]	Classify illnesses of sunflower leaves	Data gathered by the researchers using Google images	Hybrid model with MoileNet and Vgg16	89.2% accuracy
[22]	Insect and disease identification in rice crops	A total of 3355 photos of artificial rice leaves were gathered from Kaggle, representing real data. In total, 200 pictures were gathered from rice fields in Gujranwala, Pakistan	Vgg16, Vgg19, ResNet50, ResNet50V2 and ResNet101V2	75.0% accuracy for artificial dataset and 86.799% accuracy for real dataset
[23]	Collected about 96,000 photos of both healthy and diseased plant leaves	Acquired about 96,000 photos of healthy and diseased plant leaves	CNN	94% accuracy
[24]	Predicting crop diseases	The PlantVillage Dataset has 54,306 photos of leaves and can diagnose 38 various illnesses	DNN	99.24% accuracy
[25]	Taxonomy of plant health	1918 images	CNN	99.58% of accuracy
[26]	Taxonomy of tomato plant illnesses	PlantVillage dataset	Random Forest	97% accuracy
[27]	Detecting tomato plant illnesses	PlantVillage dataset	ResNet-50+ SeNet	96.81% accuracy
[28]	Detection of tomato plant diseases	PlantVillage dataset	Lightweight CBAM attention module-based ResNet20	99.69% accuracy
[29]	Identification of health and many diseased tomato leaf pictures	18,161 tomato leaf photos from the PlantVillage dataset	EfficientNet-B4 and EfficientNet-B7	99.89% accuracy

significant real-time data through affordable, energy-efficient sensors and equipment [30]. Ground surveys in agriculture are expensive and typically cover limited areas or a subset of sample farms. In contrast, remote sensing offers a cost-effective means of agricultural monitoring [31]. These devices generate real-time data and communicate over the open internet, necessitating measures to mitigate security and data privacy risks. Data privacy is crucial for farmers, not just large organisations [32]. One approach to address the data security challenge is for data owners to collaborate in developing and utilising shared ML models while ensuring that all training data remains local. Federated learning (FL), an emerging ML technique, facilitates this collaborative approach.

The FL framework was introduced by the Google team in October 2016 [33], with the initial aggregation method, FedAvg, unveiled the same year [34]. FL represents a form of ML wherein multiple clients or users collaborate to train a model while maintaining decentralised control over their respective data [35]. This approach minimises the risk of privacy breaches [36]. FL facilitates the training of ML models across numerous decentralised clients, utilising only local data samples, thus obviating the need to share data with other clients or a central server [36].

FL proves to be a highly efficient method, particularly when training with real data sourced from distributed devices. It offers advantages over training on centralised data stored in a single location, especially when dealing with sensitive data. Additionally, processing capacity is distributed across clients, necessitating more resources for learning about clients in the federated network. FL boasts numerous applications [37], with particular significance in the realm of agriculture. However, there has yet to be a systematic review of FL applications in agriculture. FL systems can adopt both centralised and decentralised architectures [34], with the only distinction between these two being in client-server communication yet yielding identical results. Notably, all evaluated FL applications in agriculture, as outlined in Table 12.2, employ a centralised architecture.

Table 12.2 Applications of federate learning in agriculture

Reference	Problem description	Dataset	Algorithms	Number of clients
[38]	Crop yield prediction	Soybean yield dataset includes soil components, weather and crop data	FedAvg	3
[39]	Intrusion detection	ToN-IoT dataset	FedGRU	2
[40]	Facilitating data exchange throughout the agri-food supply chain	Crop yield prediction datasets based on both image (remote sensing) and tabular data (weather and soil)	FedBN extends FedAvg	5
[41]	Leaf disease prediction	PlantVillage	FedAvg	4
[42]	Diagnoses of plant diseases	Sensor technologies and IoT platforms are used in conjunction with unmanned aerial vehicles (UAVs). Pest images	FedAvg	4

(Continues)

Table 12.2 (Continued)

Reference	Problem description	Dataset	Algorithms	Number of clients
[43]	Improving productivity	Soil and crop data	Amendable multi-function sensor control method (AMFSC)	47
[44]	Managing agricultural productivity	The photos from the real-world soybean iron deficiency chlorosis (IDC) dataset	To discover the ideal solution, a combined FL framework for edge-assisted Internet of Agriculture Things (Edge IoAT) was proposed, along with a greedy algorithm	10
[45]	Data privacy in intelligent agriculture	The dataset has independent variables of temperature, humidity, pH and rain	Federated averaging model	6
[46]	To avoid expensive data storage and communications expenses, imbalanced and insufficient data from orchards, a diversity of pests and illnesses, and difficult detection situations using typical powered by the cloud deep learning	There are 445 images of orchard apples, with just 152 including five illnesses	FedAvg	6

12.2.4 Split learning

Split learning (SL), also known as split neural networks (SplitNN), addresses the challenge of training a deep neural network on diverse data entities. While SL inherently enhances privacy protection, no studies have explored its compatibility with common privacy-preserving techniques such as encryption. In [47], researchers investigated SL's effectiveness in handling sensitive time-series data and found it insufficient to ensure the privacy of training one-dimensional (1D) CNN algorithms. Specifically, they demonstrated that raw data could be extracted from the activation patterns of the intermediate split layer. To mitigate privacy risks, the authors proposed two remedial approaches: incorporating hidden layers and implementing varying levels of privacy. However, their findings indicate that neither strategy effectively prevents privacy leakage across all activation pathways of SL. Additionally, both approaches lead to a reduction in the joint algorithm's accuracy.

The SL approach trains a model with a client and a server to protect the user's personal information. If the SL model is not employed, and the client wishes to train his or her data on the server's DL model, raw data is given to the server. In this situation, if sensitive information is stolen, the attacker has access to the original data, resulting in a violation of privacy. To avoid such a privacy infringement, the client pre-trains data using specific models before sending it to the server. It is impossible to reconstruct the original data without knowing the model's structure and weights.

SL has yet to conduct significant model-related research, and its performance is not superior to that of a generic model. Because the CNN is primarily explored, it is important to examine voice data, image data and string data. Furthermore, when an attacker steals data between a client and a server, a white box attack is achievable provided the attacker is familiar with the client's model structure as well as some input and output data. In this instance, because model reconstruction attacks are feasible, the original data may be calculated. Even in this circumstance, we need a mechanism to defend ourselves from attacks.

12.3 The emergence of splitfed learning

SL emerges as a more suitable option for edge computing compared to FL due to its compatibility with resource-constrained environments and model privacy considerations. However, the sequential client-side model synchronisation in SL results in increased latency during model training/testing, posing challenges in scenarios requiring rapid updates with constantly evolving information. This scenario is particularly common in industries such as healthcare and finance, where distributed environments demand swift model updates for tasks like real-time anomaly detection and fraud recognition. To leverage the advantages of both FL and SL, SFL is introduced.

SFL, an innovative method to FL, has emerged as a possible answer to the issues confronting smart agriculture. Unlike classic FL methods, which rely on centralised data aggregation, SFL distributes model training over several edge devices while maintaining data privacy and reducing communication cost.

SFL, an innovative method to FL, has emerged as a possible answer to the issues confronting smart agriculture. Unlike classic FL methods, which rely on centralised data aggregation, SFL distributes model training over several edge devices while maintaining data privacy and reducing communication cost. We illustrate the architecture in three major tiers as shown in Figure 12.2. These levels are connected by two connecting layers. We separate them into two sublayers because the connectivity layers connect distinct layers using different technologies. The interconnection layer is the key layer of smart agricultural architecture because it acts as a bridge between all of the layers, allowing all of them to function in sync with one another.

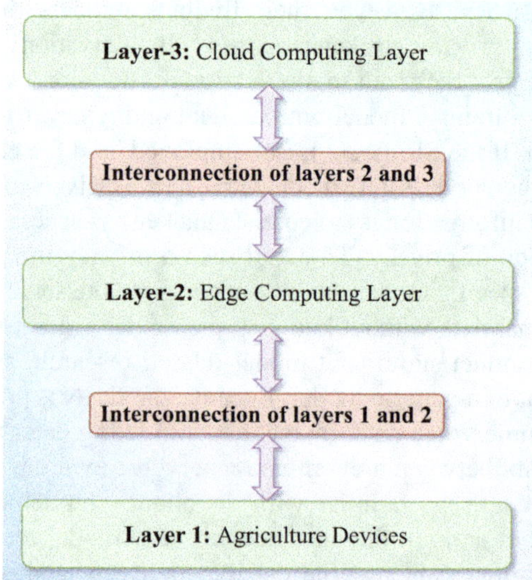

Figure 12.2 Layered architecture of smart agriculture

At the core of the smart agricultural architecture lies layer 1: Agricultural devices. This foundational layer encompasses an array of components, including sensors strategically positioned across agricultural landscapes, animal enclosures, greenhouses, hydroponic systems, tagged livestock, unmanned aerial vehicles (UAVs), agricultural robots, automated fencing systems and tractors. These devices, also referred to as scattered source nodes, continuously monitor environmental conditions, collect real-time data and transmit it to the gateway node located in the subsequent layer through the interconnecting infrastructure, which primarily comprises a Wireless Sensor Network (WSN). For instance, in a rice crop field, data is gathered by subterranean soil sensors, on-UAV sensors and cameras, which are then relayed to the edge for processing.

Layer 2: Edge computing comprises multiple edge nodes, the quantity of which is determined by the configuration of the smart agriculture system. Here, data collected at Layer 1 undergoes analysis, filtering and encryption. Previously, due to resource limitations at the edge layer, predictive and remedial tasks were typically delegated to the subsequent layer. However, recent advancements in hardware and AI at the edge initiatives have enabled trained ML models to perform predictive analytics and offer solutions at this layer. Nonetheless, for resource-intensive tasks or those lacking urgency, both prediction and inference can still be carried out in the following layer. For example, in a livestock farm, if a cow strays from its designated area or requires milking, the edge computing layer takes appropriate actions and alerts the farmer accordingly. Various hardware boards are employed as edge devices [48]. Some notable examples include the Arduino UNO used in [49] for agricultural control and monitoring, the Raspberry Pi utilised in a hydroponic system [50], the ESP8266 for connecting intelligent agricultural functionalities and regulating environmental factors [51], the ESP32 for intelligent irrigation systems [52], the Intel Edison chip for vertical farming facilities [53] and the BeagleBone for tracking agrochemical methods [54].

SFL is a technique that allows for collaborative model training while protecting the confidentiality and safety of highly sensitive agricultural information. Here's a full description of how SFL functions in this context:

1. *Decentralised data ownership:* In agriculture, data is frequently acquired from multiple sources, including farms, weather stations, soil sensors and satellite images. Each source of data maintains ownership and control over its own data. This is critical for safeguarding privacy and keeping sensitive information under the control of the organisation that obtained it.
2. *Split model architecture:* SL separates the ML model into two components: the client and the server. The client part is located on the data storage devices such as individual farms or agricultural research institutions. The server element is in charge of coordinating the entire training process and managing the global model parameters.
3. *Model training process:*
   - Random parameters are used to initialise the global model on the server.
   - During training, the server provides a copy of the current model parameters to each client.
   - The client trains the model privately on its own data, without sharing it with the server or other clients.
   - After training, the client only communicates model changes (gradients) to the server, not raw data.
   - The server collects model updates from all clients to update global parameters.
   - The method iterates until the global model converges or a predetermined stopping threshold is met.
4. *Privacy and security measures:*
   - Clients retain control over their raw data, eliminating the need to communicate sensitive information with a central server or other clients. This assures data protection and compliance with legislation such as the General Data Protection Regulation (GDPR).
   - Encrypt communication between clients and servers to prevent data eavesdropping and unauthorised access.
   - Integrating techniques like differential privacy and FL can improve model security and prevent privacy breaches.

Layer 3: Cloud computing, the virtual tier, typically resides in data centres accessible globally via the Internet [55]. The immense volume of data generated by sensors or cameras on agricultural farms requires processing, analysis and storage. Historically, the cloud has served as the primary hub for decision-making, analysis and storage of vast datasets [56–58]. Its robust computational capabilities allow for the execution of diverse complex tasks within reasonable timeframes. Nevertheless, cloud computing faces constraints necessitating the exploration of alternative computing models. Challenges such as latency, high bandwidth internet requirements and concerns regarding data security and privacy impede the real-time monitoring and management of smart agriculture systems.

Interconnecting layers: These layers facilitate communication between different tiers of the architecture. The interconnecting layer-1 receives physical attribute data from Layer 1 and forwards it to Layer 2. Similarly, interconnecting layer-2 transmits processed data from Layer 2 to Layer 3. Various communication networks are utilised in this layer, depending on the geographical area that needs to be interconnected. For instance, when data transitions from the farming device layer to the edge computing layer, short-range communication technologies such as Wi-Fi, Bluetooth and near-field communication (NFC) are commonly employed, along with ZigBee, Z-Wave and radio frequency identification (RFID). For larger farms requiring communication over longer distances, LoRaWAN, Narrowband IoT (NB-IoT) and SigFox are preferred due to their low energy consumption and long-distance transmission capabilities. Conversely, in small farms situated in rural areas with limited network speed and minimal battery consumption, Z-Wave serves as a suitable alternative. Bluetooth Low Energy (BLE) has been utilised for soil and air monitoring systems, as well as water management systems [59], while ZigBee has found application in agriculture system management [60]. RFID technology is widely adopted in the smart agriculture industry [48–51,61,62] and LoRa has been employed for water management applications [52].

When processed data transitions from the edge computing layer to the cloud layer, cellular technologies such as ground penetrating radar services (GPRS), long-term evolution (LTE), 3G/4G and 5G come into play. The latest 5G technology offers minimal latency, high reliability, extensive coverage, rapid data rates, and introduces new frequency bands [63], presenting significant opportunities for the advancement of smart agriculture. GPRS has been utilised for irrigation systems [58], while ongoing initiatives are leveraging the capabilities of 5G [52,53]. The evolution of cellular technology continues with the development of 6G, promising even faster speeds compared to current mobile networks. This next generation of cellular technology will foster flexible decentralised models, accelerating advancements in various fields such as blockchain, edge computing and AI, consequently driving the expansion of intelligent agriculture.

12.4 Data privacy-enhancing techniques

The impact of intelligent sensors on agricultural growth is due to recent technical advancements. Leveraging spatial, temporal and visual data can enhance prediction accuracy in agriculture. Sensors gathering information on crops, weather, soil and other agricultural variables at regular intervals contribute to improving ML effectiveness, enhancing agricultural production management and promoting sustainability.

Both FL and SL address the systemic privacy problems associated with classic centralised learning by including a data reduction principle into their architecture. FL implements this notion by combining information from various data records into a targeted model update. In contrast, SL restricts the amount of information acquired about each data record by only releasing the crushed data. SFL, notably in agriculture data privacy and security, is an approach that enables collaborative model training while protecting the privacy and security of sensitive agricultural data. Section 12.3 explains how SFL works in this setting.

For the sake of simplicity, we'll presume that the number of measures shown is proportionate to the privacy risk, or the amount of private information leaked. We should stress that this simplified privacy model is far from a rigorous assessment of data privacy, but it can serve as a starting point with space for improvement in the future. Let us suppose a client that gives *n_c* confidential data samples for training. Assume the FL model has N_w elements (i.e., $w \in R^{N_w}$) and the eliminated layer size/smashed data dimensionality is *d* (i.e., h_{w^c} (x) $\in R^d$). In every training round, the typical number of features disclosed to the server per training data is $\frac{N_w}{n_c}$ for FL and *d* for SL. As the algorithm's update is for a predetermined number of aspects, FL advantages over having a large amount of local data for training to mitigate the privacy issue. Conversely, the amount of detail disclosed for each sample of data in SL is independent of the local information size and only depends on the client's model design (i.e., cut layer size). Layers that are deeper create more compact depictions (i.e., smaller *d*); hence, the client model in SL frequently has to choose between computational effectiveness and privacy. FL seems to be better suited for circumstances when every client has a sufficient number of data samples ($\geq$ *N_w*/*d*), but prevents customers with smaller quantities of data from contributing due to the greater privacy risk. However, because recent deep neural networks are growing at a rapid pace, the bare minimum needed data amount can soon scale up to a level that is extremely difficult to attain in practice. For instance, if we develop a ResNet-50 model [32] with over 23 million parameters that can be trained to recognise input images and split after implementing an adaptive pooling layer (corresponding to a cut layer size of 512), we would need at least 44,000 data records from each site to justify the use of FL. Such a threshold is frequently impossible for agriculture applications, as a client with only a few farm records might nonetheless make a significant contribution to the model. In these instances, SL becomes more appealing since it is more adaptable and may offer the same level of privacy advantages irrespective of the extent of the client's local data.

Risk of inversion: When compared with FL, SL reduces the dangers of modelling/gradient inversion attacks [54–67] by limiting model accessibility for both parties. Learning can be defined as applying a function *F* to the client's private data X to get a result M. In FL, although the raw data X is not directly available to the server, the computation of the function *F* and the outcome M have been disclosed to the server, allowing the server to deduce the inverse function in order to infer X from M and violate client's privacy. In contrast, SL divides the function *F* (i.e., the model) into two parts: the client's F_C and the server's F_S, with only the latter being revealed to the server. As a result, the SL server only knows imprecise knowledge of the computing process, making it difficult to determine the exact inverse function.

12.5 Applications, benefits and challenges

Smart agriculture uses SFL for a variety of purposes, including crop production prediction, disease detection, soil monitoring and precision farming. Farmers can gain meaningful information in near real time by leveraging the collective intelligence of dispersed edge devices, allowing for proactive decision-making and resource optimisation.

SFL has considerable benefits for agricultural privacy, collaborative learning and scalability. By decentralising data ownership, allowing for collaborative model training and supporting scalable deployments across remote devices, SL enables stakeholders to gain useful insights from agricultural data while protecting individual privacy and data security.

1. *Privacy preservation:* SL allows data owners to maintain control over their data while participating in global model training, ensuring privacy preservation. This method has various benefits for privacy preservation.
   - *Data ownership:* Data ownership is decentralised, allowing farms, research organisations and agricultural enterprises to maintain control of their own data.
   - *Data confidentiality:* Raw data is not shared with a central server or third parties during model training. Instead, only model changes are communicated, ensuring the security of sensitive data.
   - *Compliance:* SFL helps organisations comply with privacy standards like GDPR, Health Insurance Portability and Accountability Act (HIPAA) and industry-specific rules by minimising data exposure and ensuring data owners have control over its use.
2. *Collaborative learning:* Multiple stakeholders can collaborate without sharing sensitive data, leading to more robust and accurate models. The collaborative nature of SL offers several benefits:
   - Knowledge sharing: Stakeholders can pool their resources and expertise to train a more accurate and robust global model without the need to share raw data.
   - Model diversity: By leveraging diverse datasets from different sources, SL helps mitigate bias and improve the generalisation capabilities of the model.
   - Domain expertise: Collaboration among domain experts, data scientists and researchers enables the development of specialised models tailored to specific agricultural challenges such as crop prediction, pest detection or soil analysis.

3. *Scalability:* SL enables scalable model training across several remote devices or data sources. This scalability feature has various benefits:
   - SL, which distributes the training workload over many devices or data sources, speeds up model training and optimises resource utilisation.
   - SL can be used in edge computing contexts, where data processing and model inference are closer to the source. This lowers latency and bandwidth utilisation, making it ideal for real-time applications in precision agriculture or IoT-based farming systems.
   - SL is adaptable to many infrastructures, including as cloud-based platforms, on-premises servers and edge devices, making it suited for agricultural deployments.

Despite its potential, SFL faces numerous problems. Addressing communication overhead, model synchronisation and security concerns is critical to the effective use of SFL in agriculture. Implementing robust communication protocols, optimising model synchronisation procedures and prioritising data privacy and security measures can help stakeholders overcome hurdles and realise the full potential of SL for agricultural applications.

1. *Communication overhead:* Aggregating model updates from several clients might result in communication overhead and latency. Several things contribute to communication issues:
   - Limited bandwidth can cause delays and communication overhead when updating models via network connections.
   - Large-scale agricultural deployments may face unacceptably expensive data transmission costs via cellular networks or satellite linkages.
   - High network congestion can cause delays in model synchronisation and updates.
2. *Model synchronisation:* Maintaining consistent model versions across clients and servers can be problematic, particularly in dynamic situations. Model synchronisation poses several significant challenges:
   - Asynchronous updates: Clients may update their local models at various times, resulting in an inconsistent global model state. When asynchronous updates are not correctly managed, they might cause divergence or stale model parameters.
   - Network latency and communication delays can disrupt synchronisation, making it challenging to maintain consistent model versions across remote devices.
   - Conflict resolution: Maintaining global model integrity requires resolving conflicts and guaranteeing data consistency when several model modifications occur at the same time.
3. *Security concerns:* Proper installation and communication protocols are crucial for preventing system breaches or attacks. Several security risks must be addressed.

- Data privacy: Protecting sensitive agricultural data from unauthorised access or publication is crucial. Encryption mechanisms, access limits and data anonymisation procedures all serve to protect data privacy and confidentiality.
- Authentication and authorisation: Implementing strong authentication techniques ensures that only authorised people and devices may access the system and take part in model training. Role-based access restrictions and secure authentication mechanisms help to prevent unauthorised access to critical resources.
- Encryption systems like SSL/TLS, HTTPS and VPNs ensure secure data transmission between clients and servers, preventing eavesdropping, tampering and interception.
- Proactive monitoring, intrusion detection systems and anomaly detection techniques can identify and mitigate security threats like malicious assaults, insider threats or data breaches.
- Compliance with regulatory frameworks, including GDPR, HIPAA and industry-specific standards, ensures data privacy and security procedures comply with legal obligations.

To address these issues, researchers, practitioners and policymakers must collaborate together across disciplines to create strong algorithms and frameworks specific to agricultural ecosystems.

12.6 Conclusion and future directions

In conclusion, SFL presents a compelling solution for collaborative model training in agricultural data analysis, prioritising data privacy and security. By decentralising data ownership and facilitating collaborative learning, it empowers stakeholders to contribute to global model development while safeguarding sensitive information. While promising, challenges such as communication overhead, model synchronisation and security concerns must be addressed. Future directions include optimising communication protocols, enhancing model synchronisation mechanisms and advancing security measures. Additionally, exploring integration with edge computing and fostering interdisciplinary collaboration can further improve the application of SFL in agriculture, fostering innovation for sustainable farming practices and food security.

References

[1] Bhojwani, Y., Singh, R., Reddy, R., and Perumal, B. Crop selection and IoT-based monitoring system for precision agriculture. In *The 2020 International Conference on Emerging Trends in Information Technology and Engineering (ic-ETITE)* (pp. 1–11). Piscataway, NJ: IEEE, 2020.

[2] Park, J., Moon, A., Lee, E., and Kim, S. Understanding IoT climate data based predictive model for outdoor smart farm. In *International Conference*

on Information and Communication Technology Convergence (ICTC), Piscataway, NJ: IEEE, 2021.

[3] Rahman, M. T., Mahmud, S., Li, Y., and Rahman, M. A. IoT based smart farming system to reduce manpower, wastage of time and natural resources in both traditional and urban mega farming. In *4th International Conference on Advanced Electronic Materials, Computers and Software Engineering (AEMCSE)*, Piscataway, NJ: IEEE, 2021.

[4] Li, Q., Wen, Z., Wu, Z., *et al.* A survey on federated learning systems: Vision, hype and reality for data privacy and protection. *IEEE Transactions on Knowledge and Data Engineering* 2021, 35(4), 3347–3366.

[5] Li, T., Sahu, A. K., Talwalkar, A., and Smith, V. Federated learning: Challenges, methods, and future directions. *IEEE Signal Processing Magazine*, 2020, 37(3), 50–60.

[6] Wang, S., Tuor, T., Salonidis, T., *et al.* Adaptive federated learning in resource constrained edge computing systems. *IEEE Journal on Selected Areas in Communications*, 2019, 37(6), 1205–1221.

[7] Ji, S., Pan, S., Long, G., Li, X., Jiang, J., and Huang, Z. Learning private neural language modeling with attentive aggregation. In *International Joint Conference on Neural Networks*, Budapest, Hungary, 14–19 July 2019.

[8] Zalik, K.R. and Zalik, M. A review of federated learning in agriculture. *Sensors*, 2023, 23, 9566. https://doi.org/10.3390/s23239566.

[9] Varghese, R. and Sharma, S. Affordable smart farming using IoT and machine learning. In *2018 Second International Conference on Intelligent Computing and Control Systems (ICICCS)*, 2018 (pp. 645–650).

[10] Condran, S., Bewong, M., Islam, M. Z., Maphosa, L., and Zheng, L. Machine learning in precision agriculture: A survey on trends, applications and evaluations over two decades. *IEEE Access*, 2022, 10, 73786–73803.

[11] Mekonnen, Y., Namuduri, S., Burton, L., Sarwat, A., and Bhansali, S. Machine learning techniques in wireless sensor network based precision agriculture. *Journal of the Electrochemical Society*, 2019, 167(3), 037522.

[12] Reddy, K. S. P., Roopa, Y. M., Ln, K. R., and Nandan, N. S. IoT-based smart agriculture using machine learning. In *2020 Second International Conference on Inventive Research in Computing Applications (CIRCA)*, 2020, July (pp. 130–134). Piscataway, NJ: IEEE.

[13] Bhanu, K. N., Jasmine, H. J., and Mahadevaswamy, H. S. Machine learning implementation in IoT-based intelligent system for agriculture. In *2020 International Conference for Emerging Technology (INCET)*, 2020, June (pp. 1–5). Piscataway, NJ: IEEE.

[14] Reshma, R., Sathiyavathi, V., Sindhu, T., Selvakumar, K., and SaiRamesh, L. IoT-based classification techniques for soil content analysis and crop yield prediction. In *2020 Fourth International Conference on I-SMAC (IoT in Social, Mobile, Analytics, and Cloud)(I-SMAC)*, 2020, October (p. 156160). Piscataway, NJ: IEEE.

[15] Sharma, A., Jain, A., Gupta, P., and Chowdary, V. Machine learning applications for precision agriculture: A comprehensive review. *IEEE Access*, 2020, 9, 4843–4873.

[16] Dahane, A., Benameur, R., Kechar, B., and Benyamina, A. An IoT based smart farming system using machine learning. In *2020 International Symposium on Networks, Computers and Communications (ISNCC)*, 2020 (pp. 1–6).

[17] Liu, Q., Yan, Q., Tian, J., and Yuan, K. Key technologies and applications in intelligent agriculture. *Journal of Physics: Conference Series*, 2021, 1757, 012059.

[18] Dargan, S., Kumar, M., Ayyagari, M.R., and Kumar, G. A survey of deep learning and its applications: A new paradigm to machine learning. *Archives of Computational Methods in Engineering*, 2020, 27, 1071–1092.

[19] LeCun, Y., Bengio, Y., and Hinton, G. Deep learning. *Nature*, 2015, 521, 436–444.

[20] Jihen, A., Bouaziz, B., and Algergawy, A. A deep learning-based approach for banana leaf diseases classification. In *Datenbanksysteme für Business, Technologie und Web (BTW 2017)-Workshopband*, Gesellschaft für Informatik e.V.: Bonn, Germany, 2017.

[21] Sirohi, A. and Malik, A. A hybrid model for the classification of sunflower diseases using deep learning. In *2nd International Conference on Intelligent Engineering and Management (ICIEM)*, London, UK, 2021, pp. 58–62.

[22] Burhan, S.A., Minhas, S., Tariq, A., and Hassan, M.N. Comparative study of deep learning algorithms for disease and pest detection in rice crops. In *Proceedings of the 2020 12th International Conference on Electronics, Computers and Artificial Intelligence (ECAI)*, Bucharest, Romania, 25–27 June 2020.

[23] Ahmed, A.A. and Reddy, G.H. A mobile-based system for detecting plant leaf diseases using deep learning. *AgriEngineering*, 2021, 3, 478–493.

[24] Pallagani, V., Khandelwal, V., Chandra, B., Udutalapally, V., Das, D., and Mohanty, S.P. DCrop: A deep-learning-based framework for accurate prediction of diseases of crops in smart agriculture. In *Proceedings of the 2019 IEEE International Symposium on Smart Electronic Systems (iSES) (Formerly iNiS)*, Rourkela, India, 16–18 December 2019.

[25] Abdullahi, H.S., Sheriff, R., and Mahieddine, F. Convolution neural network in precision agriculture for plant image recognition and classification. In *Proceedings of the 2017 Seventh International Conference on Innovative Computing Technology (INTECH)*, Luton, UK, 16–18 August 2017 (pp. 1–3).

[26] Khan, S. and Narvekar, M. Novel fusion of color balancing and superpixel based approach for detection of tomato plant diseases in natural complex environment. *Journal of King Saud University – Computer and Information Sciences*, 2022, 34(6), 3506–3516.

[27] Zhao, S., Peng, Y., Liu, J., and Wu, S. Tomato leaf disease diagnosis based on improved convolution neural network by attention module. *Agriculture*, 2021, 11, 651.

[28] Bhujel, A., Kim, N.E., Arulmozhi, E., Basak, J.K., and Kim, H.T. A lightweight attention-based convolutional neural networks for tomato leaf disease classification. *Agriculture*, 2022, 12, 228.

[29] Chowdhury, M.E., Rahman, T., Khandakar, A., *et al.* Automatic and reliable leaf disease detection using deep learning techniques. *AgriEngineering*, 2021, 3, 294–312.

[30] Morais, R., Mendes, J., Silva, R., Silva, N., Sousa, J.J., and Peres, E. A versatile, low-power and low-cost IoT device for field data gathering in precision agriculture practices. *Agriculture*, 2021, 11, 619.

[31] Wójtowicz, M., Wójtowicz, A., and Piekarczyk, J. Application of remote sensing methods in agriculture. *Communications in Biometry and Crop Science*, 2016, 11, 31–50.

[32] Kaur, J., Hazrati Fard, S.M., Amiri-Zarandi, M., and Dara, R. Protecting farmers' data privacy and confidentiality: Recommendations and considerations. *Frontiers in Sustainable Food Systems*, 2022, 6, 903230.

[33] Konĕcnỳ, J., McMahan, H.B., Yu, F.X., Richtárik, P., Suresh, A.T., and Bacon, D. Federated learning: Strategies for improving communication efficiency. arXiv:1610.05492v2, https://doi.org/10.48550/arXiv.1610.05492.

[34] McMahan, H.B., Moore, E., Ramage, D., Hampson, S., and Arcas, B.A. Communication efficient learning of deep networks from decentralized data. In *Proceedings of the 20th International Conference on Artificial Intelligence and Statistics (AISTATS) 2017*, Fort Lauderdale, FL, USA, 2017 (pp. 1273–1282).

[35] Kairouz, P., McMahan, H.B., Avent, B. *et al.* Advances and open problems in federated learning. *Foundations and Trends® in Machine Learning*, 2021, 14, 1–210.

[36] Mothukuri, V., Parizi, R.M., Pouriyeh, S., Huang, Y., Dehghantanha, A., and Srivastava, G. A survey on security and privacy of federated learning. *Future Generation Computer Systems*, 2021, 115, 619–640.

[37] Rahman, K.J., Ahmed, F., Akhter, N., *et al.* Challenges, applications and design aspects of federated learning: A survey. *IEEE Access*, 2021, 9, 124682–124700.

[38] Manoj, T., Makkithaya, K., and Narendra, V.G. A federated learning-based crop yield prediction for agricultural production risk management. In *Proceedings of the 2022 IEEE Delhi Section Conference (DELCON)*, New Delhi, India, 11–13 February 2022 (pp. 1–7).

[39] Kumar, P., Gupta, G.P., and Tripathi, R. PEFL: Deep privacy-encoding-based federated learning framework for smart agriculture. *IEEE Micro*, 2021, 42, 33–40.

[40] Durrant, A., Markovic, M., Matthews, D., May, D., Enright, J., and Leontidis, G. The role of cross-silo federated learning in facilitating data

sharing in the agri-food sector. *Computers and Electronics in Agriculture*, 2022, 193, 106648.

[41] Antico, T. M., Moreira L. F. R., and Moreira, R. Evaluating the potential of federated learning for maize leaf disease prediction. In *Anais do XIX Encontro Nacional de Inteligência Artificial e Computacional*, Sociedade Brasileira de Computação: Porto Alegre, Brazil, 2022 (pp. 282–293).

[42] Khan, F. S., Khan, S., Mohd M. N. H., *et al.* Federated learning-based UAVs for the diagnosis of plant diseases. In *Proceedings of the 2022 International Conference on Engineering and Emerging Technologies (ICEET)*, Kuala Lumpur, Malaysia, 27–28 October 2022 (pp. 1–6).

[43] Abu-Khadrah, A., Ali, A. M., and Jarrah, M. An amendable multi-function control method using federated learning for smart sensors in agricultural production improvements. *ACM Transactions on Sensor Networks*, 2023. https://doi.org/10.1145/3582011.

[44] Yu, C., Shen, S., Zhang, K., Zhao, H., and Shi, Y. Energy-aware device scheduling for joint federated learning in edge-assisted internet of agriculture things. In *Proceedings of the 2022 IEEE Wireless Communications and Networking Conference (WCNC)*, Austin, TX, USA, 10–13 April 2022 (pp. 1140–1145).

[45] Idoje, G., Dagiuklas, T., and Iqbal, M. Federated learning: Crop classification in a smart farm decentralised network. *Smart Agricultural Technology*, 2023, 5, 100277.

[46] Deng, F., Mao, W., Zeng, Z., Zeng, H., and Wei, B. Multiple diseases and pests detection based on federated learning and improved faster R-CNN. *IEEE Transactions on Instrumentation and Measurement*, 2022, 71, 3523811.

[47] Abuadbba, S., Kim, K., Kim, M. *et al.* Can we use split learning on 1D CNN models for privacy preserving training? In *Proceedings of the 15th ACM Asia Conference on Computer and Communications Security*, 2020 (pp. 305–318).

[48] Friha, O., Ferrag, M. A., Shu, L., Maglaras, L., and Wang, X. Internet of things for the future of smart agriculture: A comprehensive survey of emerging technologies. *IEEE/CAA Journal of Automatica Sinica*, 2021, 8(4), 718–752. doi:10.1109/JAS.2021.1003925.

[49] Hamrita, T.K. and Hoffacker, E.C. Development of a "smart" wireless soil monitoring sensor prototype using RFID technology. *Applied Engineering in Agriculture*, 2005, 21(1), 139–143.

[50] Peets, S., Gasparin, C.P., Blackburn, D.W.K., and Godwin, R.J. RFID tags for identifying and verifying agrochemicals in food traceability systems. *Precision Agriculture*, 2009, 10(5), 382–394.

[51] Ruiz-Garcia, L. and Lunadei, L. The role of RFID in agriculture: Applications, limitations and challenges. *Computers and Electronics in Agriculture*, 2011, 79(1), 42–50.

[52] Sjolander, A.J., Thomasson, J.A., Sui, R., and Ge, Y. Wireless tracking of cotton modules. Part 2: Automatic machine identification and system testing. *Computers and Electronics in Agriculture*, 2011, 75(1), 34–43.

[53] Vellidis, G., Tucker, M., Perry, C., Kvien, C., and Bednarz, C. A real-time wireless smart sensor array for scheduling irrigation. *Computers and Electronics in Agriculture*, 2008, 61(1), 44–50.

[54] Zhao, W., Lin, S., Han, J., Xu, R., and Hou, L. Design and implementation of smart irrigation system based on LoRa. In *Proceedings of IEEE Globecom Workshops (GC Wkshps)*, Piscataway, NJ: IEEE, 2017 (pp. 1–6).

[55] Gupta, M., Abdelsalam, M., Khorsandroo, S., and Mittal, S. Security and privacy in smart farming: Challenges and opportunities. *IEEE Access*, 2020, 8, 34564–34584. doi:10.1109/ACCESS.2020.2975142.

[56] Khattab, A., Abdelgawad, A., and Yelmarthi, K. Design and implementation of a cloud-based IoT scheme for precision agriculture. In *Proceedings of 28th International Conference on Microelectronics (ICM)*, 2016 (pp. 201–204). doi:10.1109/ICM.2016.7847850.

[57] Na, A. and Isaac, W. Developing a human-centric agricultural model in the IoT environment. In *Proceedings of International Conference on Internet of Things and Applications (IOTA)*, 2016 (pp. 292–297). doi:10.1109/IOTA.2016.7562740.

[58] López-Riquelme, J.A., Pavón-Pulido, N., Navarro-Hellín, H., Soto-Valles, F., and Torres-Sánchez, R. A software architecture based on FIWARE cloud for precision agriculture. *Agricultural Water Management*, 2017, 183, 123–135.

[59] Hernández-Rojas, D.L., Fernández-Caramés, T.M., Fraga-Lamas, P., and Escudero, C.J. Design and practical evaluation of a family of lightweight protocols for heterogeneous sensing through BLE beacons in IoT telemetry applications. *Sensors*, 2018, 18(1), 57.

[60] Mafuta, M., Zennaro, M., Bagula, A., Ault, G., Gombachika, H., and Chadza, T. Successful deployment of a wireless sensor network for precision agriculture in Malawi. *International Journal of Distributed Sensor Networks*, 2013, 9(5), 150703.

[61] Kodali, R.K., Jain, V., and Karagwal, S. IoT based smart greenhouse. In *Proceedings of IEEE Region 10 Humanitarian Technology Conference (R10-HTC)*, Piscataway, NJ: IEEE, 2016 (pp. 1–6).

[62] Komarek, A.M., De Pinto, A., Smith, V.H. A review of types of risks in agriculture: What we know and what we need to know. *Agricultural Systems*, 2020, 178, 102738.

[63] Villa-Henriksen, A., Edwards, G.T.C., Pesonen, L.A., Green, O., and Gron Sorensen, C.A. Internet of things in arable farming: Implementation, applications, challenges and potential. *Biosystems Engineering*, 2020, 191, 60–84.

[64] Faraci, G., Raciti, A., Rizzo, S., and Schembra, G. A 5G platform for unmanned aerial monitoring in rural areas: Design and performance issues. In *Proceedings of 4th IEEE Conference on Network Softwarization and Workshops (NetSoft)*. Piscataway, NJ: IEEE, 2018 (pp. 237–241).

[65] Alahmadi, A., Alwajeeh, T., Mohanan, V., and Budiarto, R. Wireless sensor network with always best connection for internet of farming. In Mohanan, V., Budiarto, R., and Aldmour, I. (eds), *Powering the Internet of Things with 5G Networks*. Hershey, PA: IGI Global, 2018 (pp. 176–201).

[66] Fredrikson M., Jha S., and Ristenpart T. Model inversion attacks that exploit confidence information and basic countermeasures. In *Proceedings of the 22nd ACM SIGSAC Conference on Computer and Communications Security*, 2015 (pp. 1322–1333).

[67] Ferrández-Pastor, F.J., García-Chamizo, J.M., Nieto-Hidalgo, M., Mora-Pascual, J., and Mora-Martínez, J. Developing ubiquitous sensor network platform using internet of things: Application in precision agriculture. *Sensors*, 2016, 16(7), 1141.

Chapter 13

A case study on splitfed learning implementation

B.C. Soundarya[1], *Gururaj H.L.*[1], *Geetabai S. Hukkeri*[2] *and Hong Lin*[3]

In this study, we examine the application of splitfed learning, a unique distributed machine learning paradigm that combines the advantages of federated learning and split learning. We explore the practical elements of implementing splitfed learning in real-world settings and evaluate its efficacy with respect to model performance, computational efficiency and data privacy. We describe in detail the communication protocols, data partitioning techniques and system architecture that enable smooth cooperation between central servers and dispersed clients. This work sheds light on the actual issues and solutions related to adopting splitfed learning, opening the path for its use in privacy-sensitive and resource-constrained settings.

Keywords: Case studies; Federated learning; Split learning; Splitfed learning; Machine learning

13.1 Introduction

Federated learning (FL) and split learning (SL) are two prevalent methods of distributed machine learning (ML). In both approaches, users train and test ML models without sharing raw data. However, SL offers enhanced model privacy compared to FL due to its separation of the ML model architecture between clients and the server. Furthermore, because of the split model, SL is a superior choice for resource-constrained applications. However, because of the relay-based training across numerous clients, SL performs slower than FL. In this regard, this chapter introduces splitfed learning (SFL), a new technique that incorporates the two approaches while eliminating the resulting drawbacks, as well as a refined structural configuration including distinct security and PixelDP to improve data privacy and model its durability. Splitfed is useful in resource-constrained contexts when

[1]Department of Information Technology, Manipal Institute of Technology Bengaluru, Manipal Academy of Higher Education, India
[2]Department of Computer Science and Engineering, Manipal Institute of Technology Bengaluru, Manipal Academy of Higher Education, India
[3]Department of Computer and Mathematical Sciences, University of Houston-Downtown, Houston, TX, USA

complete model training and installation are impossible and quick model training time is necessary to continually update the global model on an evolving dataset. These contexts define different domains.

The rapid development of networking as well as distributed computing capabilities has empowered the widespread mobilisation of the Internet of Things (IoTs), in which a massive amount of data created by various consumer devices is analysed via multiple applications. As datasets grow higher, training ML algorithms increasingly necessitates sharing model parameter optimisation across numerous processors. 'Distributed Collaborative Machine Learning (DCML)' is notable in view of its inherent data privacy advantages [1]. In contrast to traditional methods that centralise and access data, DCML allows ML without needing data transfer from data custodians to untrusted parties. In this approach, analysts do not have access to the source data; rather, the ML framework is sent to the data custodian for computing. Additionally, DCML supports processing across multiple platforms or servers, including distributed devices.

FL [2,3] and SL [4] are among the most widely adopted DCML techniques. FL involves training an ML model on distributed clients using their local data, followed by aggregating these locally trained models on a server to create a global model. A key advantage of FL is its ability to facilitate parallel and thus efficient training of ML models across numerous users. Nonetheless, FL necessitates clients to run the entire ML model, which can be prohibitively expensive for resource-constrained devices, such as those in the IoT. This challenge is particularly pronounced with deep learning models. Additionally, during training, both the server and clients have full access to local and global models, posing a privacy risk from the model point of view.

SL was developed to address these concerns. SL partitions the entire ML model into smaller segments, which are then trained separately on a server and on distributed clients using local data. This approach significantly minimises the computing burden over resource-constrained devices by allowing them to train only a portion of the network, unlike in FL where the entire model must be executed [5]. Additionally, in SL, clients do not have access to the centralised model and vice versa, enhancing privacy and architecture as shown in Figure 13.1.

Despite the benefits of SL, there is a major drawback. Since only one client communicate with the server at one time in relay-based training with SL, this leads to idle resources on the clients side, resulting in a considerable increase in training overhead with numerous clients. This chapter offers a unique architecture termed splitfed learning to overcome these difficulties in FL and SL. SFL explores the benefits of FL and SL at the time of emphasising data privacy and model strength.

FL is an ML method that allows model training to take place across several decentralised edge devices like mobile phones, IoT devices without the need for raw data to be exchanged. Traditional ML algorithms necessitate centralising all data in one area, which can be inconvenient for huge datasets, as well as inefficient due to excessive communication costs. In place of transmitting source data to a centralised server, FL sends the model to each device [6]. On the basis of its local data, each device computes an update to the model. These modifications are subsequently sent back to the central server, which collects them to update the

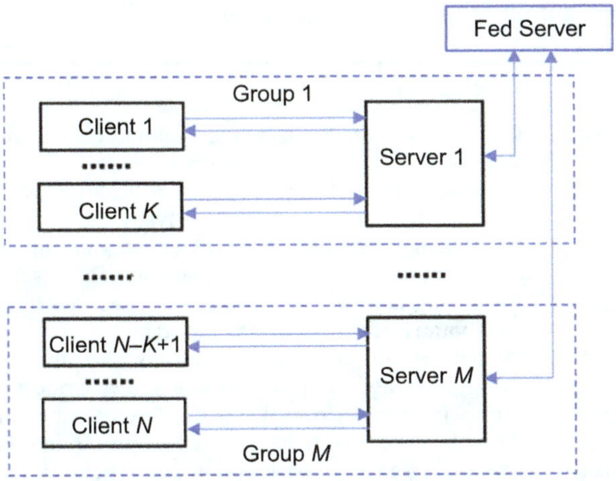

Figure 13.1 Splitfed learning process

centralised model. This procedure is performed iteratively, allowing the model to benefit from the pooled understanding of all participating machines while keeping data localised and private. FL has numerous applications, including finance, healthcare and IoT, where data confidentiality and communication effectiveness are critical. SL is a strategy that falls under the umbrella of FL. SL splits the model into two components: a local half and a centralised part. The local component processes the raw data on the user and device and only transmits the intermediate approximations to the server-side component. The server-side component merges these preliminary representations from different systems and maintains the global model. This framework guarantees the raw data is never sent outside of the local device, protecting user privacy. SL is especially beneficial when the raw data is confidential or too vast to be efficiently delivered over the network. SL lowers communication bandwidth utilisation and privacy concerns by keeping data processing local and communicating only essential information. Table 13.1 shows the applications of split federated learning.

13.2 Case studies

13.2.1 A case study of Google Gboard

FL is an H2020/EU and FP7/2007–2013 ML technique. Rather than using centralised servers, it seeks to enable model training on end-user devices. Better privacy is a benefit of this approach because only model updates are transmitted and raw data is never removed from the user's device [7,8]. It will be necessary to implement novel and creative solutions to deal with data handling and model training given the increase in privacy concerns in all AI applications. Since FL

Table 13.1 Application of split federated learning

Application	Description
Healthcare diagnostics	Collaborative diagnosis across multiple healthcare institutions while preserving patient privacy
Financial fraud detection	Detection of fraudulent activities in banking and finance sectors while maintaining data privacy
Smart city infrastructure	Optimising urban infrastructure and services by analysing data from various sources without sharing sensitive information
Telecommunications networks	Enhancing network performance and security through collaborative analysis of network data
Agricultural yield prediction	Predicting crop yields and optimising agricultural practices by aggregating data from farms while ensuring farmer privacy
Autonomous vehicles	Improving the performance and safety of autonomous vehicles through collaborative learning while preserving data privacy
Energy management systems	Optimising energy consumption and efficiency in smart grids by analysing data from multiple sources without revealing sensitive information
Industrial IoT applications	Enhancing efficiency and productivity in industrial settings by analysing sensor data from distributed devices while maintaining data privacy

disentangles the training process from each user's local device, it has emerged as a workable alternative.

This chapter presents the architecture, implementation and results of our investigation on implementing FL to Google Gboard. With capabilities like next-word prediction, Google's Gboard is a virtual keyboard app that requires a large amount of data to train on. Historically, these models would have required sending large volumes of user data to the cloud, which could have been risky for privacy.

13.2.1.1 Implementation of federated learning in Gboard

Architecture

1. Local training: Typing data is saved on users' devices. Using this data, the Gboard app trains a local ML model.
2. Model updates: Model updates (such as gradients or model weights) are periodically transmitted to a central server.
3. Aggregation: To enhance the global model, these updates from many devices are combined by the central server.
4. Distribution: The modified global model is subsequently sent back to the consumers' devices.

Privacy mechanisms

1. Data minimisation: The central server receives only model updates—raw data is not shared with it.

2. Secure aggregation: To make sure that a single update cannot be linked to a particular user, methods like secure aggregation are employed [9].
3. Differential privacy: To prevent the identification of specific users, differential privacy is implemented by adding noise to the model updates.

Accuracy improvement

- Baseline comparison: When compared to the baseline model trained on centralised data, the FL-based Gboard performed significantly better in terms of next-word prediction accuracy.
- Performance metrics: Evaluation parameters such as precision, recall and F1-score showed significant improvement.

Privacy impact

- User trust: According to surveys, users' faith in Gboard's privacy protections has increased.
- Data security: There were no reports of data leakage, proving FL's strength in safeguarding privacy [10].

Benefits

1. Enhanced privacy: FL substantially decreases the danger of data breaches by storing user data on the device.
2. Model personalisation: The local training feature enables models to be better tailored to specific user actions.
3. Scalability: Without the need to centralise enormous volumes of data, the federated approach scales effectively over millions of devices.

Challenges

1. Communication overhead: Regular model updates may result in higher latency and communication costs.
2. Device heterogeneity: Model consistency may be impacted by variations in data quality and device performance.
3. Computational load: Increased computing demands on consumer devices may have an effect on battery life and device performance.

13.2.1.2 Future work

Further research should focus on improving FL algorithms to better handle device heterogeneity, streamlining communication protocols to minimise overhead and creating more effective privacy-preserving methods.

13.2.2 Enhancing financial fraud detection with splitfed learning

Financial fraud is a widespread problem that costs organisations billions of dollars every year. Large-scale data sharing is necessary for fraud detection using traditional centralised ML techniques, which poses serious privacy and security issues. This work provides a thorough case study on the hybrid strategy known as SFL, which combines SL and FL approaches, and is applied to the detection of financial fraud [11]. Financial companies can jointly train ML models using this technique without jeopardising sensitive consumer data.

It investigates the creative use of SFL, particularly for fraud detection, in the financial industry. SFL improves model accuracy while maintaining data privacy by fusing the advantages of SL and FL. This method meets the criteria for collaboration and data sensitivity by spreading the learning process among several financial institutions and a central server. The case study illustrates how SFL may increase fraud detection rates while protecting the privacy of client data.

13.2.2.1 Financial fraud detection

The process of detecting fraudulent and illegal activity in financial systems is known as financial fraud detection. With their ability to analyse transaction patterns and identify abnormalities suggestive of fraud, ML models are indispensable. FL enables decentralised data centralisation while enabling different entities to cooperatively train a common model. Every participant trains the model locally, sharing with the central server only the updates from the model, which the server aggregates. In SL, a neural network is divided into multiple components. While the subsequent layers are trained on a central server, the first few layers are trained on the client side. Using this method guarantees that raw data never leaves the client's location.

FL and SL are combined in SFL. Clients locally train the model's first few layers, then communicate the intermediate results to a central server to finish the training process. By combining updates from several clients, the server allows for collaborative model improvement while protecting client anonymity.

13.2.2.2 Implementation of splitfed learning in financial fraud detection

Architecture

1. Local training: Using its local data, each participating financial institution trains a neural network's first few layers. Transaction histories, customer profiles and past fraud incidents are all included in this data.
2. Intermediate outputs: A central server receives the intermediate representations that come from the locally trained layers. Because these representations don't include raw data, privacy is preserved.
3. Central training: The central server trains the neural network's remaining layers using intermediate outputs that are sent to it from various universities.
4. Model aggregation: The server improves the global model by combining updates from all universities.
5. Model distribution: The enhanced global model is returned to each institution so that local data can be used to further modify it.

Privacy and security mechanisms

1. Data anonymisation: By using intermediate representations rather than raw data, the possibility of sensitive information being revealed is decreased.
2. Secure aggregation: To make sure that intermediate representations cannot be linked back to particular organisations or clients, methods like secure multi-party computing and homomorphic encryption are used.
3. Differential privacy: The identification of specific transactions or consumers is hindered by the addition of noise to model updates.

Accuracy improvement

- Baseline comparison: When compared to conventional models, the SFL-based fraud detection system has shown a notable improvement in identifying fraudulent activity.
- Performance metrics: Precision, recall, F1-score and area under the ROC curve (AUC-ROC) were used to evaluate the model. The splitfed model outperformed the baseline in all metrics.

Privacy impact

- User trust: Participating institutions' surveys revealed a rise in confidence in the system's capacity to safeguard private client information.
- Data security: The resilience of the privacy-preserving procedures was validated by the fact that no data breaches or leaks were reported during installation.

Benefits

1. Enhanced privacy: SFL enables high privacy levels by only sharing intermediate representations and retaining raw data within the institutions.
2. Improved detection rates: SFL's collaborative nature facilitates more thorough model training by leveraging a variety of data sources to improve the accuracy of fraud detection.
3. Scalability: With its ability to grow across multiple institutions, the splitfed method makes widespread adoption possible without requiring a great deal of data centralisation.

Challenges

1. Communication overhead: Frequent exchanges of intermediate representations may result in higher latency and expense of communication.
2. Model consistency: The consistency and effectiveness of the global model may be impacted by differences in data quality and device performance throughout institutions.
3. Computational load: The need for on-site training raises the computational demands on institutional resources, potentially affecting daily operations.

Future work

Future directions for research include developing more effective privacy-preserving methods, improving SFL algorithms to manage data heterogeneity and optimising communication protocols to minimise overhead. Furthermore, investigating SFL's connection with cutting-edge technologies like blockchain could improve data security and trust even further.

13.2.3 Healthcare data collaboration with split learning

SL is used in the medical field with an emphasis on cooperative medical image analysis [12]. SL protects patient privacy while allowing several healthcare facilities to train ML models together. This case study shows how SL may leverage dispersed data to increase diagnostic accuracy while protecting private patient data.

Large volumes of data are produced by the healthcare sector, which is essential for building reliable ML models for diagnostic applications. However, data sharing across institutions is restricted by privacy laws and the delicate nature of patient data. This problem is tackled by SL, which permits cooperative model training without requiring the sharing of raw data. This work showcases the potential of SL to improve data privacy and diagnostic accuracy through a case study of its use to medical image analysis [13].

13.2.3.1 Medical image analysis

Medical image analysis is the process of interpreting medical images, such as CT, MRI and X-rays, using ML models [14]. To attain high accuracy, these models need a large amount of training data, which is frequently dispersed across several universities. SL divides a neural network into multiple pieces in order to facilitate collaborative ML. The intermediate representations are delivered to a central server, which completes the remaining training, after the initial segments are trained on local data at client locations. By guaranteeing that raw data never leaves the client's location, this technique improves security and privacy [15].

13.2.3.2 Implementation of split learning in healthcare data collaboration

Architecture

1. Data distribution: At participating healthcare facilities, patient medical pictures and related records are kept locally.
2. Local training: Every organisation uses its own local data to train the neural network's first few layers. Preprocessing photos and removing pertinent elements are part of this.
3. Intermediate representations: A central server receives the intermediate representations, or feature maps, produced by the locally trained models.
4. Central training: All participating universities send intermediate representations to the central server, which uses them to finish training the neural network's final layers.
5. Model aggregation: After updating the global model with the results compiled by the central server, it is distributed back to the institutions.
6. Refinement: Organisations can improve overall performance by using the global model to better hone their local models.

Privacy and security mechanisms

1. Data anonymisation: Only intermediate representations, which do not contain raw patient data, are shared with the central server.
2. Encryption: To guard against interception and guarantee data security, intermediate representations are encrypted while being transmitted.
3. Secure aggregation: Secure multi-party computation (SMPC), one of the advanced cryptographic algorithms used, makes sure that data from several institutions can be combined without disclosing individual contributions.

Accuracy improvement

- Baseline comparison: When comparing the SL-based model to models trained on data from specific universities, there was a noticeable gain in diagnosis accuracy.
- Performance metrics: Measures including area under the receiver operating characteristic curve (AUC-ROC), sensitivity and specificity showed improved performance when using the SL strategy.

Privacy impact

- User trust: Participating universities' surveys revealed strong levels of confidence in SL's ability to protect privacy [16].
- Data security: The absence of any documented data breaches or leaks confirms the efficacy of the security measures put in place.

Benefits

1. Enhanced privacy: SL guarantees high levels of data privacy by retaining raw data inside the institutions and only sharing intermediate representations.
2. Improved diagnostic accuracy: SL's collaborative nature makes it possible to train models more thoroughly by leveraging a variety of datasets to improve diagnostic accuracy.
3. Scalability: SL is widely adopted without requiring significant data centralisation because it scales well across numerous universities.

Challenges

1. Communication overhead: Increased delay and communication costs may result from the frequent exchange of intermediate representations [17].
2. Model consistency: The consistency and effectiveness of the global model may be impacted by differences in data quality and device performance throughout institutions.
3. Computational load: The need for on-site training raises the computational demands on institutional resources, potentially affecting daily operations.

Future work

Future topics for study include developing more effective privacy-preserving methods, improving SL algorithms to handle data heterogeneity better and optimising communication protocols to minimise overhead. Furthermore, investigating how SL might be integrated with cutting-edge technologies like blockchain could improve data security and trust even further.

13.2.4 Enhancing Siri and Dictation with federated learning

It looks at how Apple uses FL to improve the functionality of Dictation and Siri on iOS devices. Apple is able to increase the precision of its natural language processing (NLP) and speech recognition models while maintaining user privacy because to FL [18,19]. Apple exhibits a noteworthy progression in the implementation of privacy-preserving AI by decentralising the training process and retaining user data on their devices.

NLP and voice recognition are essential for the operation of Dictation and other iOS device functions, as well as virtual assistants like Siri [5]. Large volumes of data are needed for traditional ML models, and these data are usually centralised on servers, which poses serious privacy issues. This study describes how Apple uses FL to improve the precision and effectiveness of Dictation and Siri while addressing these issues.

13.2.4.1 Voice recognition and NLP

While NLP allows machines to comprehend and analyse human language, voice recognition entails turning spoken language into text. Virtual assistants need these technologies in order to carry out activities like delivering messages, making reminders and giving information.

13.2.4.2 Federated learning

Federated learning (FL) is an innovative machine learning paradigm designed to address data privacy and efficiency concerns by decentralizing the learning process. This is particularly relevant in the context of the Internet of Things (IoT), where numerous devices generate and collect vast amounts of data.

13.2.4.3 Implementation of federated learning in Siri and Dictation

Architecture

1. Local training: Using interaction data from the user, each iOS device locally trains a portion of the NLP and voice recognition models. Text inputs and voice recordings are included in this.
2. Model updates: Apple's central servers get locally trained model changes (e.g., gradients or model weights) on a regular basis. These updates are encrypted to protect data during transmission.
3. Aggregation: To enhance the global model, the central server compiles the updates it has received from multiple devices. The use of aggregation techniques, such as secure aggregation, makes sure that individual updates cannot be linked to particular users.
4. Model distribution: A continuous learning loop is created by returning the modified global model to users' devices, where fresh local data is used to improve it even more.

Privacy mechanisms

1. Data minimisation: User interactions are kept private because only model updates—not raw data—are transferred to the central server.
2. Secure aggregation: Sophisticated cryptographic methods guarantee that the server securely aggregates the changes without having direct access to the individual updates.
3. Differential privacy: Apple also protects the anonymity of user data by adding noise to the model updates through the use of differential privacy approaches.

Accuracy improvement

- Baseline comparison: Compared to baseline models trained on centralised data, the adoption of FL greatly increased the accuracy of Siri's voice recognition and Dictation functions.
- Performance metrics: Metrics like accuracy in NLP tasks and word error rate (WER) for speech recognition demonstrated a noticeable improvement. Better performance in comprehending and reacting to voice commands was reported by users.

Privacy impact

- User trust: After learning that their data is still stored on Apple devices, consumers' confidence in the company's privacy policies grew, according to surveys and user reviews.
- Data security: The efficacy of the privacy protections was validated by the fact that no data breaches were reported during the installation.

Benefits

1. Enhanced privacy: FL dramatically lowers the danger of data breaches and unauthorised access by storing user data on-device.
2. Improved model performance: The model's capacity to comprehend and analyse various accents, languages and usage patterns is enhanced by the collaborative training approach, which enables the model to learn from a wide range of user interactions.
3. Scalability: FL allows for ongoing enhancements to Dictation and Siri without requiring a significant amount of data centralisation because it scales effectively across millions of iOS devices.

Challenges

1. Communication overhead: Model updates that are transmitted often may result in higher communication costs and possible delay problems. This can be mitigated by optimising data compression methods and update frequency.
2. Device heterogeneity: The global model's performance and consistency could be impacted by differences in data quality and device performance

among various iOS devices. Strategies for managing this heterogeneity are crucial.

3. Computational load: Computational resources are needed for local training, which may affect battery life and device performance. To solve this, effective power management techniques and training algorithms are required.

Future work

Future topics for research include developing more effective privacy-preserving methods, improving FL algorithms to better handle device heterogeneity and optimising communication protocols to minimise overhead. Investigating how FL might be integrated with other cutting-edge AI technologies, including edge computing, could improve user experience and performance even more.

13.3 Conclusion

These privacy-preserving ML techniques have the potential to revolutionise a variety of domains, as demonstrated by the case studies that explore FL, SL and SFL. Every case study emphasises the distinct benefits and difficulties that come with implementing these technologies in practical settings, highlighting their relevance and significance.

References

[1] Kairouz, P., McMahan, H. B., Avent, B., *et al.*, Advances and open problems in federated learning. *Foundations and Trends® in Machine Learning* 14 (2021): 1–210.

[2] Gupta, O., and Raskar, R. Distributed learning of deep neural network over multiple agents. *Journal of Network and Computer Applications* 116 (2018): 1–8.

[3] Joshi, P., Thapa, C., Camtepe, S., *et al.*, Splitfed learning without client-side synchronization: Analyzing client-side split network portion size to overall performance. In *Collaborative European Research Conference*, Cork, Ireland, pp. 1–7, 2021.

[4] Sumanth, N., Heggere, S. B., Prathiksha, E., Shetty, C. A., Nikhitha, H. R. H., and Soundarya, B. C. Federated learning: Healthcare, security, challenges, and threats (2023). https://www.researchsquare.com/article/rs-3223963/v1.

[5] Konečný, J., McMahan, H. B., Yu, F. X., Richtárik, P., Suresh, A. T., and Bacon, D. Federated learning: Strategies for improving communication efficiency. arXiv:1610.05492 (2016), https://doi.org/10.48550/arXiv.1610.05492.

[6] Bonawitz, K., Eichner, H., Grieskamp, W., *et al.*, Towards federated learning at scale: System design. *Proceedings of Machine Learning and Systems* 1 (2019): 374–388.

[7] McMahan, B., Moore, E., Ramage, D., Hampson, S., and Aguera y Arcas, B. Communication-efficient learning of deep networks from decentralized data.

In *Proceedings of the 20th International Conference on Artificial Intelligence and Statistics (AISTATS) 2017*, Fort Lauderdale, FL, USA, 2017 (pp. 1273–1282).

[8] Vepakomma, P., Gupta, O., Swedish, T., and Raskar, R. Split learning for health: Distributed deep learning without sharing raw patient data. arXiv:1812.00564 (2018), https://doi.org/10.48550/arXiv.1812.00564.

[9] Li, X., Gu, Y., Dvornek, N., Staib, L. H., Ventola, P., and Duncan, J. S. Multi-site fMRI analysis using privacy-preserving federated learning and domain adaptation. In *International Conference on Medical Image Computing and Computer-Assisted Intervention (MICCAI)* (2019).

[10] Sheller, M. J., Edwards, B., Reina, G. A., *et al.*, Federated learning in medicine: Facilitating multi-institutional collaborations without sharing patient data. *Scientific Reports* 10(1) (2020): 1–12.

[11] Chen, Y., Luo, X., and Weng, J. SplitFed: When federated learning meets split learning. In *Proceedings of the 2021 IEEE International Conference on Communications (ICC)* (2021).

[12] Mammen, P. M. Federated learning: Opportunities and challenges. arXiv:2101.05428 (2021), https://doi.org/10.48550/arXiv.2101.05428.

[13] Rieke, N., Hancox, J., Li, W., *et al.*, The future of digital health with federated learning. *NPJ Digital Medicine* 3, no. 1 (2020): 1–7.

[14] Xu, J., Glicksberg, B. S., Su, C., Walker, P., Bian, J., and Wang, F. Federated learning for healthcare informatics. *Journal of Healthcare Informatics Research* 5 (2021): 1–19.

[15] Goetz, J., Malik, K., Bui, D., Moon, S., Liu, H., and Kumar, A. Active federated learning. arXiv:1909.12641 (2019), https://doi.org/10.48550/arXiv.1909.12641.

[16] Ren, Y. and Jiang, B. W. On the granulation and irregular variation of red supergiants. *The Astrophysical Journal* 898 (2020): 24.

[17] Smith, T. M., and Chiang, F. Federated learning for financial services. *Journal of Banking and Finance Technology* 4(2) (2020): 85–102.

[18] Hard, A., Rao, K., Mathews, R., *et al.*, Federated learning for mobile keyboard prediction. arXiv:1811.03604 (2018), https://doi.org/10.48550/arXiv.1811.03604.

[19] Li, L., Fan, Y., Tse, M., and Lin, K.-Y. A review of applications in federated learning. *Computers and Industrial Engineering* 149 (2020): 106854.

Chapter 14
Conclusion and future directions

Geetabai S. Hukkeri[1], *Gururaj H.L.*[2], *N.Z. Jhanjhi*[3] *and Hong Lin*[4]

14.1 Introduction

This book, *Split Federated Learning for Secure IoT Applications: Concepts, frameworks, applications and case studies*, offers a thorough examination of split federated learning (splitfed) and its transformational potential across diverse IoT domains. Each chapter has covered a variety of topics, ranging from fundamental principles regarding processing techniques to particular applications in industries such as healthcare, education, transportation and agriculture. In this final chapter, we summarise the significant conclusions from previous chapters, address the remaining questions and suggest future research possibilities that will move this discipline ahead.

14.1.1 Recapitulation of key insights

The introduction chapter presented the fundamentals of federated learning, split learning and the unified splitfed learning strategy. The first chapter examined the use of splitfed learning in IoT and big data contexts, emphasising its ability to handle and analyse large amounts of data that is distributed efficiently. Chapter 2 described blockchain-driven splitfed learning for improving data security in IoT environments, emphasising the relationship in blockchain technology and splitfed systems.

Chapters 3 and 4 looked into splitfed learning approaches in natural language processing (NLP) and recommendation systems, correspondingly. These chapters showed how splitfed can be customised to increase efficiency and privacy in certain applications. Chapter 5 explored programmable intelligent surface-inspired splitfed learning for wireless applications, highlighting advances in wireless communication technology.

[1]Department of Computer Science and Engineering, Manipal Institute of Technology Bengaluru, Manipal Academy of Higher Education, India
[2]Department of Information Technology, Manipal Institute of Technology Bengaluru, Manipal Academy of Higher Education, India
[3]School of Computer Science (SCS), Taylor's University, Malaysia
[4]Department of Computer and Mathematical Sciences, University of Houston-Downtown, Houston, TX, USA

The following series of chapters concentrate on domain-specific applications. Chapter 6 examined improving computing performance in healthcare using federated learning, demonstrating the practical implications of medical data analysis and treatment. Chapter 7 investigated splitfed learning for multimodal emotion recognition, opening up new possibilities in interacting with computers and mental health monitoring. Chapter 8 looked at splitfed learning used for educational data analysis, emphasising its ability to personalise the learning process to enhance educational results.

Chapters 9–12 expanded the discussion to include smart city applications, smart transportation, smart grids and smart agriculture. Several chapters provide thorough case studies and illustrations of how splitfed learning may improve system efficiency, sustainability and user happiness in several essential infrastructure areas.

The case studies in Chapter 13 demonstrate the versatility and broad effect of split federated learning spanning multiple sectors. These applications achieve enhanced speed, security and efficiency by utilising splitfed learning despite tackling the unique demands and problems of each area.

14.2 Open questions

14.2.1 Scalability and efficiency

As splitfed learning evolves, one important difficulty remains: scaling effectively for vast IoT networks with numerous units. Future studies must address computing and communication overheads, as well as create effective load balancing mechanisms, to assure large-scale deployment feasibility.

14.2.2 Standardisation and interoperability

The lack of standardised protocols might hinder the effortless incorporation of various IoT devices into splitfed learning platforms. Future work should concentrate on defining universal standards and assuring interoperability across multiple platforms and devices in order to encourage wider adoption.

14.2.3 Energy efficiency

IoT devices frequently function with very few electrical resources, making energy economy an important consideration. Future research should focus on optimising splitfed learning architectures to reduce energy consumption, hence increasing the lifespan of IoT devices involved in the process of learning.

14.2.4 Dynamic network conditions

IoT networks experience communication and bandwidth variations. Creating sophisticated algorithms that can retain learning results in these dynamic environments is critical for real-world applications. Future research should focus on improving splitfed learning systems' robustness to network unpredictability.

14.3 Future directions

14.3.1 Advanced machine learning techniques

Advanced machine learning approaches, such as reinforcement learning and generative adversarial networks, can be combined with splitfed learning to improve the system's learning capacity and robustness. Future studies should look into these linkages to see how far splitfed learning can take us.

14.3.2 Cross-domain applications

Splitfed learning's use beyond the subjects being addressed domains, like monitoring the environment, smart cities and handling emergencies, can open up new research and development opportunities. These cross-domain applications may gain advantages from splitfed learning's adaptability and security characteristics.

14.3.3 Edge and fog computing integration

Integrating splitfed learning with fog computing and edge computing approaches can help minimise delay and enhance real-time data processing. This integration is critical for crucial IoT applications and should be the primary objective of future research efforts.

14.3.4 Hybrid models

Creating hybrid models that integrate splitfed learning with other federated or centralised learning methodologies can provide flexibility while optimising performance for specific use cases. Future research should look into the potential advantages and disadvantages of such hybrid models.

14.3.5 User-centric approaches

Relying on user-focused techniques that involve end-users in the learning process can increase personalisation and happiness, especially in applications such as recommendation systems and health care. Future study should look into ways to increase user involvement and contribution in splitfed learning systems.

14.3.6 Ethical and societal implications

Understanding the ethical and societal consequences of splitfed learning, such as data ownership, permission and the digital gap, can help guarantee that technical improvements are consistent with social principles and ethical standards. New studies should address these critical issues in order to promote responsible innovation.

14.4 Concluding remarks

In conclusion, this book provides a thorough review of splitfed learning and its potential uses, raising various issues and prospects for future research. Ongoing collaboration across disciplines will be critical in addressing these obstacles and realising the full promise of splitfed learning for secure IoT applications. The path ahead looks to be both fascinating and consequential, leading the way for smarter, more reliable and efficient IoT ecosystems. As we look ahead, it is apparent that splitfed learning will play an important part in designing the coming generations of IoT applications and systems.

Glossary of terms

5G	5G is the fifth generation of wireless networking technology, offering faster data rates, less latency, and more reliable connections.
6G	6G is the next generation of mobile network technology, projected to provide even greater speed, dependability, and connectivity than 5G.
Activation Values	The outputs of nodes (neurons) in a neural network after applying an activation function, which determines the neuron's role in the model.
Active RIS	Reconfigurable Intelligent Surfaces with active components for improved signal reflection and control.
AdaBoost	AdaBoost is a boosting technique that combines numerous weak classifiers to form a strong classifier, hence enhancing prediction accuracy.
Adaptive Learning Systems	Educational systems that adapt content and instructional methods to the individual learner's needs and abilities.
Advanced data analysis	Advanced data analysis refers to techniques and tools for processing and interpreting complicated data in order to derive relevant insights.
Advanced Encryption Methods	Sophisticated approaches for protecting data by encrypting it in a way that only authorised parties can decode.
Analog RIS	RIS that manipulates continuous signals rather than transforming them to digital.
Application-Specific Integrated Circuits (ASICs)	Integrated circuits that are customised for specific applications and provide excellent performance and efficiency.
Application-specific RIS	A wireless communication system that has been built and optimised for certain use cases or applications.
Applied Research	Applied research is research that aims to solve practical problems using scientific methods and understanding.
Arduino UNO	Arduino UNO is a popular microcontroller board for creating digital gadgets and interactive things that can perceive and control the physical world.

Artificial Intelligence (AI)	AI is the simulation of human intellect in machines that accomplish activities that normally require human cognition.
Autoencoders	Autoencoders are neural networks used for unsupervised learning, generally to reduce dimensionality or denoize data.
Back Propagation	It is a neural network approach that updates weights by propagating errors backwards from the output layer to the input layer.
Bandwidth Limitations	The limited capacity for data transmission via a network, which can hinder or limit data exchange and communication.
Big Data	Extremely massive and complicated data collections that necessitate cutting-edge storage, processing, and analysis techniques.
Biosensors	Biosensors are devices that detect and transform biological information into signals for study and monitoring.
Centralized Learning Approaches	Machine learning techniques that collect and process data in a centralised location or server.
Cloud Computing	Cloud computing is the supply of information technology services over the internet, providing immediate availability of computational resources as well as information storage.
Cognitive Capability	Cognitive capability is the mental ability to learn, think, reason, and remember.
Collaborative Model Training	Collaborative Model Training: Multiple parties train a machine learning model without revealing their own data, which improves privacy and leverages distributed resources.
Communication Overhead	It is the additional communication necessary in distributed systems to coordinate tasks and transmit data, which can have an impact on performance and efficiency.
Complex Programmable Logic Devices (CPLDs)	These are programmable logic devices with lower complexity and capabilities than FPGAs that are utilised for simpler jobs.
Computational Models of Emotion	These are theoretical and practical models that use computational approaches to simulate and analyse human emotions.
Configuration Mode	The state in which a system is configured or changed before switching to operating mode.
Control Logic	Control logic is the set of rules and algorithms that regulate how a system operates and makes decisions.
Convolutional Neural Networks (CNNs)	CNNs are a type of deep neural network that is mostly used for image and video processing. They use convolutional layers to automatically and adaptively learn spatial hierarchies of data.

Coordinated Averaging Model	In distributed learning, many local models are averaged to produce a global model update.
Cost-Effectiveness	A measure of how efficiently resources is used to achieve a goal while balancing costs and benefits.
CRMKL Model	A variation of MKL that uses collaborative learning to improve prediction accuracy.
Customer Behavior Analysis	It is the study of consumer behavior patterns using data analytics to influence corporate strategies and improve the customer experience.
Cybersecurity	Cybersecurity refers to the measures and policies that secure systems, networks, and data from cyberattacks and unauthorised access.
Data Anonymization	Data anonymization refers to techniques for protecting privacy by eliminating or obscuring personal identifiers from data sets.
Data Confidentiality	Data confidentiality entails ensuring that information is only accessible to those who are authorised to have access, hence preventing unauthorised disclosure.
Data Custodian	A person or organization in charge of managing and safeguarding data, assuring its correctness, privacy, and adherence to legislation.
Data Encryption	It is the process of transforming plaintext data into a coded format (ciphertext) in order to prevent unauthorised access and ensure data security.
Data Privacy and Security	Practices and technologies used to secure data from unauthorised access and breaches while maintaining user privacy.
Database for Emotion Analysis Using Physiological Signals (DEAP)	It is a database of physiological and visual signals used in emotion recognition research.
Data-Driven Approaches	Techniques and methods that use data analysis to guide decisions and actions.
Decentralized Collaborative Learning Recommendation (DCLR)	It is a recommendation strategy in which many entities train a recommendation model together without centralizing their data.
Decision Tree	A supervised learning technique used for classification and regression tasks that makes judgements by dividing data into branches.
Deep Belief Networks (DBNs)	These are a sort of generative neural network made up of several layers of stochastic, latent variables.
Deep Learning (DL)	It is a subset of machine learning that employs neural networks with several layers capable of learning complex patterns.
Deep Neural Networks (DNN)	Deep Neural Networks (DNN) are neural networks that have numerous hidden layers and can learn complex patterns and representations from data.

Deep Radial Basis Functions	Advanced neural network designs that employ radial basis functions in deep learning applications.
Deep Radial Basis Techniques	Advanced neural network methods that activate radial basis functions, commonly utilised in function approximation and pattern recognition.
Deep Reinforcement Learning (DRL)	Deep Reinforcement Learning (DRL) is a type of machine learning that uses both deep learning and reinforcement learning to address complicated decision-making problems.
Development Tools	Development Tools are software and hardware tools used to design, test, and debug programmes and systems.
Differential Privacy	A strategy for protecting the privacy of individuals in a dataset by introducing noise into the data or calculations, making it harder to deduce personal information.
Digital Agriculture	Digital agriculture is the use of digital technologies in farming techniques to increase productivity, sustainability, and profitability.
Digital RIS	An RIS that converts signals to digital form, allowing for more accurate and flexible control over signal processing.
Dimensionality Reduction Techniques	Methods for reducing the number of variables in a dataset while preserving crucial information.
Distributed Collaborative Machine Learning (DCML)	Collaborative model training over numerous dispersed nodes is intended to improve learning efficiency and model performance.
Distributed Computing	Distributed computing is the utilisation of numerous computer systems connected over a network to achieve a common purpose.
Distributed Energy Resources (DERs)	Small-scale local generation devices linked to the grid, such as solar panels and wind turbines, that supply power near to the point of use.
Distribution Networks	Distribution networks transfer electricity from transmission systems to end customers at lower voltages than transmission networks.
Dropout Prediction	Using data analytics and machine learning to forecast which students are likely to drop out of a course or programme.
Drug discovery	Drug discovery is the process of identifying novel candidate treatments using computational models and data analysis.
Dynamic Reconfiguration	Dynamic reconfiguration is the capacity to change system settings or architecture in real time during operation in order to adapt to changing situations.
Early Warning Systems	Systems that detect and alert to potential threats or issues in advance, allowing for preventive action.

E-Commerce Platforms	Online markets that enable the purchase and sale of goods and services, frequently using data analytics to provide personalized experiences.
Educational Process Mining	Educational Process Mining is the use of data mining techniques on educational data to analyse and improve learning processes.
Educational Recommender Systems	Educational systems that tailor content and instructional approaches to each learner's specific needs and abilities.
Electrode Caps	Caps fitted with electrodes used for non-invasive measurement of electrical activity from the scalp, commonly in EEG.
Electrodes	Electrodes are conductive pads or devices that detect electrical activity within the body, such as brain or muscle signals.
Electroencephalogram (EEG)	A technique for measuring electrical activity in the brain, commonly utilised in medical and research settings.
Electromagnetic Waves	Electromagnetic waves are electromagnetic field waves that propagate over space. Examples include radio waves, microwaves, infrared, visible light, ultraviolet, X-rays, and gamma rays.
Electronic Health Records (EHRs)	These are digital reproductions of patients' paper charts that contain extensive health information.
Emotion Detection	Emotion detection involves identifying and analysing human emotions using a variety of data sources and algorithms.
Emotion Recognition Applications	Programmes and systems designed to identify and understand human emotions for a variety of purposes.
Emotion Recognition Technologies	Software that analyses data to identify and interpret human emotions.
Empirical Research	Empirical research is a method of collecting knowledge by direct and indirect observation or experience, which generally includes data gathering and analysis.
Energy management systems	These are systems that monitor, control, and optimise energy use in a variety of applications.
Energy Optimization	Energy optimization refers to the techniques and technologies used to minimize energy consumption and enhance efficiency in systems and buildings, which frequently incorporate AI and IoT.
Environmental Sensing and Monitoring	The use of sensors and data analysis to track and evaluate environmental conditions and changes.
Eye Pair Classifier	Eye Pair Classifier is a computer vision algorithm that detects and analyses the position and orientation of eyeballs in facial images.
Facial Image Analysis	Methods for analysing facial features and expressions for purposes such as emotion detection or biometric identification.

Feature Extraction	Feature extraction is the process of translating raw data into a set of features that may be utilised to improve the performance of machine learning models.
Federated Averaging (FedAvg)	An algorithm in federated learning that averages model updates from numerous clients to provide a global model update, thereby improving performance and privacy.
Federated Learning (FL)	A machine learning strategy in which models are trained across several decentralised devices using local data rather than transmitting it to a central server.
Federated Median	Federated Median is a federated learning aggregation method that employs the median of model parameters from several clients to improve robustness to outliers.
Feedback Mechanisms	Processes or systems that use feedback to modify or enhance performance or behaviour.
Field-Programmable Gate Arrays (FPGAs)	These are reconfigurable integrated circuits that can be programmed to execute specific jobs after they have been manufactured.
Fine-grained Emotion Identification	Fine-grained emotion identification is the careful investigation and recognition of minor changes in emotional states.
Fine-grained Emotions	Fine-grained emotions are detailed and subtle emotional states that go beyond fundamental classifications such as happy or sad.
Forward Propagation	In a neural network, input data is passed through the network layers to produce an output prediction.
Fraud Detection	Fraud detection is the application of machine learning and data analysis to detect and prevent fraudulent behaviour in financial transactions and other sectors.
Frequency Selective Surfaces (FSS)	These are structures designed to filter or control electromagnetic waves based on their frequency.
Frequency-selective RIS	RIS that can dynamically modify their response to various frequencies in order to improve communication performance.
Frontal Face Cascade Classifier	A computer vision method that detects faces in images using patterns on the frontal face.
Full-Duplex Mode	A communication mode in which data is transmitted and received simultaneously on the same channel.
Galvanic Skin Response (GSR)	A measure of skin conductivity that changes with sweat and is used to detect emotion and stress.
General Data Protection Regulation (GDPR)	A rule in EU law that governs data protection and privacy for individuals, establishing tight guidelines for the collecting, storage, and handling of personal information.

Generative Adversarial Networks (GANs) These are a type of machine learning framework in which two neural networks, a generator and a discriminator, are trained concurrently to produce realistic synthetic data.

Gradient Descent Gradient Descent is an optimisation procedure that minimises a function by iteratively moving to the steepest descent based on the function's gradient.

Gradient Updates Changes made to a neural network's weights during training based on the gradient of the loss function to reduce error.

Graph Neural Networks (GNNs) GNNs are neural networks that interact with graph topologies to capture links and dependencies between things represented as nodes and edges.

Grid Stability and Reliability The power grid's ability to provide consistent, stable, and reliable electricity without interruption.

Grid-edge Intelligence Grid-edge intelligence refers to technologies and analytics that optimise the management and operation of electrical networks.

Ground Penetrating Radar Services (GPRS) A technology that uses radar pulses to picture the subsurface, which is widely utilised in geology, archaeology, and construction.

Head-mounted Displays (HMDs) These are wearable devices that give visual output directly in front of the user's eyes, as utilised in VR and AR.

High-Performance Computing Cluster A high-performance computing cluster is a collection of interconnected computers that act as a single system to deliver large amounts of computational power.

Home Automation Home automation is the use of Internet of Things (IoT) devices to control and automate household systems such as lighting, heating, and security for convenience and efficiency.

Homomorphic Encryption An encryption method that allows computations to be conducted on ciphertexts, resulting in an encrypted output that, when decoded, corresponds to the outcome of plaintext operations.

Identity Management It refers to the processes and technology required to manage and safeguard people's digital identities, including correct authentication and authorization.

Immersive Virtual Reality VR systems that offer a highly engaging and interactive user experience, typically via head-mounted displays.

Improved Genetic Algorithm (IGA) A more advanced variant of genetic algorithms, which are optimisation approaches based on natural selection and genetic principles.

Information Gathering It is the practice of gathering data and information from many sources in order to analyse and make decisions.

Integrated RIS	RIS that are built into existing infrastructure or devices to improve functionality and performance.
Intelligent Farming	Intelligent farming is the use of AI, IoT, and data analytics to improve agricultural processes and decision-making.
Intelligent Routing Algorithms	Algorithms that find the best paths for data or vehicles, increasing efficiency and minimising travel or gearbox times.
Intelligent Systems	Intelligent systems are those that use artificial intelligence and other advanced technology to do activities that would normally require human intelligence.
Interference Mitigation	Techniques and tactics for minimising or eliminating interference in wireless communication networks.
International Business Machines (IBM)	It is a multinational technology firm that provides computer hardware, software, and consulting services.
Internet of Medical Things (IoMT)	A network of medical equipment and applications linked to healthcare IT systems, allowing for remote monitoring and management.
Internet of Things (IoT)	The Internet of Things (IoT) is a network of interconnected objects that communicate and exchange data to create smart surroundings and automated processes.
Internet Protocol (IP) Statistics	Information and metrics about the performance and application of the Internet Protocol in networking.
Interoperability	Interoperability is the capacity of different systems and organisations to collaborate and share information seamlessly.
Knowledge Distillation	Knowledge Distillation is a process in which a smaller, simpler model is trained to mimic the behaviour of a bigger, more complex model, typically to enhance efficiency.
Label Noise	Label noise refers to errors or inconsistencies in the labelling of training data that can have a negative impact on machine learning model performance and accuracy.
Learning Management Systems (LMS)	LMS are software systems that enable the administration, documenting, tracking, reporting, and delivery of educational courses and training programmes.
Line-of-Sight (LoS) Transmission	It is a type of communication in which the signal passes straight from the transmitter to the receiver with no hindrance.
Liquid crystal-based reflectors	Liquid crystal-based reflectors are reflecting surfaces that use liquid crystals to dynamically control and change their reflective qualities.

Local Model Development	The process of building machine learning models on local devices or edge nodes, which is commonly employed in federated learning.
Logistic Regression	Logistic regression is a statistical method for binary classification that calculates the likelihood of a binary outcome based on one or more predictor variables.
Machine Learning (ML)	ML is an artificial intelligence field that focuses on building algorithms that enable computers to learn from and forecast data.
Mean Absolute Error (MAE)	MAE is a statistic for assessing model accuracy that calculates the average absolute errors between predicted and actual data.
Mean Absolute Percentage Error (MAPE)	A measure of prediction accuracy of a forecasting method, calculated as the average absolute percentage error between forecasted and actual values.
Medical imaging analysis	Medical imaging analysis is the use of image processing techniques on medical pictures to aid in diagnosis and therapy planning.
Mini-Batch Gradient Descent	A hybrid of batch and stochastic gradient descent, which uses tiny batches of training data for each update to balance performance and speed.
mMTC	It is meant to accommodate a high number of devices and enable IoT connectivity in 5G networks.
Modulating Sensors	Sensors that can change their sensitivity or reaction in response to external stimuli or control signals.
Multimodal	Using various types of data or input, such as text, audio, and video for analysis.
Multiple Kernel Learning (MKL)	is a machine learning technique that uses numerous kernels to increase model performance.
Multi-sensor Mask	A mask equipped with multiple sensors to monitor the wearer's physiological signals.
Naive Bayes	A simple probabilistic classifier based on Bayes' theorem, which assumes feature independence.
Natural Language Processing (NLP)	NLP is a branch of artificial intelligence which examines the interface between machines and the language of humans, enabling machines to understand, decode, and generate language that is natural.
Nearly-Passive Reconfigurable Intelligent Surfaces	RIS technology that uses less electricity to change its features and improve wireless connection.
Non-IID (Non-Independent and Identically Distributed)	Data that is not independently and identically distributed, posing issues for machine learning models due to non-uniformity and correlation.

Online Learning Platforms	Web-based systems that offer educational information and courses, allowing for remote learning.
Operation Mode	Operation Mode is the state in which a system performs its major functions after being configured.
Parallel Processing	Parallel processing refers to the simultaneous processing of data by many processors in order to accelerate calculation and task execution.
Passive Beam Configuration	It is the use of passive devices to steer and shape an electromagnetic wave beam in the absence of active amplification.
Passive RIS	Reconfigurable Intelligent Surfaces that reflect signals without the need for active power amplification.
Pattern Recognition	Pattern Recognition is the automatic detection of patterns and regularities in data using algorithms.
Pedagogical Approach	The pedagogical approach refers to the methods and strategies utilised in teaching and education to assist learning.
Personalized Learning	Personalised learning is the process of tailoring educational content and experiences to the requirements and preferences of individual learners, frequently employing data analytics and artificial intelligence.
Personalized medicine	Personalised medicine is the practice of tailoring medical treatment to individual features, such as genetics, in order to enhance outcomes.
Personalized Recommendations	Systems that employ data and algorithms to make tailored recommendations to users, improving their experience and happiness.
Poisoning Attacks	Poisoning attacks are malicious assaults on machine learning models in which the adversary manipulates training data to degrade the model's performance.
Privacy Considerations	Policies and procedures designed to preserve individual privacy, particularly in the context of data collection and use.
Privacy protection	Efforts to secure individuals' personal information from unauthorised access and usage.
Privacy-Preserving	Privacy-preserving techniques and procedures are those that protect data privacy and security, especially in data sharing and collaborative situations.
Programming Models	Abstractions and frameworks that govern how software is written and performed, influencing both the development process and performance.
Question-Answering Systems	Artificial intelligence systems that use vast datasets and complex algorithms to respond to human-posed queries in natural language.

Radial Basis Functions (RBFs)	Radial Basis Functions (RBFs) are functions that are employed in a variety of applications, including neural networks, due to their ability to respond locally to inputs.
Radial Deep Transfer (RDT)	RDT is a machine learning technique that uses radial basis function networks with transfer learning to increase model performance.
Radio Frequency Identification (RFID)	It uses electromagnetic waves to automatically recognise and track tags affixed to things.
Random Forest	Random Forest is an ensemble learning method that builds many decision trees and aggregates their predictions to increase accuracy and reduce overfitting.
Random Projection	Random Projection is a dimensionality reduction approach that use random matrices to project high-dimensional data into lower-dimensional space.
Raspberry Pi	It is a small, low-cost computer used to learn programming and make electronics projects.
Recommendation Systems (RS)	are algorithms and systems that propose suitable things to consumers based on their preferences and previous behavior.
Reconfigurable Intelligent Surfaces (RIS)	These are surfaces having programmable features that may dynamically change their interaction with electromagnetic waves to improve wireless communication.
Recurrent Neural Networks (RNNs)	RNNs are a type of neural network built for sequential data processing, with node connections forming directed cycles that allow temporal information to be used.
Reinforcement Learning	Reinforcement Learning is a sort of machine learning in which an agent learns to make decisions by doing behaviours in an environment that maximise cumulative reward.
Renewable Energy Installations	Facilities and equipment designed to create energy from renewable sources such as solar, wind, and hydropower.
Resource Constraints in IoT Devices	IoT devices have limitations in processing power, memory, and battery life, which affect their capacity to conduct sophisticated computations.
Resource Management	Resource management is the efficient and effective use and allocation of resources like time, money, and materials.
Resource-Constrained Devices	Devices with restricted computing power, memory, and battery life, limiting their capacity to do complex computational tasks.
Root Mean Squared Error (RMSE)	RMSE is a statistic for evaluating model performance that takes the square root of the average of squared differences between forecasts and actual observations.

Scalability Issues	Difficulties related to improving a system's capacity or performance to handle increasing amounts of work or data efficiently.
Science, Technology, Engineering, and Mathematics (STEM)	An educational and professional field devoted to these four disciplines.
Secure Aggregation	A cryptographic technique used in federated learning to securely aggregate data from numerous sources while preserving individual data pieces.
Self-generating Sensors	Sensors that produce their own electricity, typically from environmental energy sources such as light or movement.
Sentiment Analysis	Sentiment analysis is the technique of employing natural language processing to discover the sentiment or emotional tone underlying a body of text.
Smart Cities	Smart Cities are urban places that use IoT, data analytics, and technology to optimise resource management, improve quality of life, and reduce environmental impact.
Smart Grid Simulation Framework	Software tools and models for simulating and analysing the performance and behaviour of smart grid systems.
Smart Grid Technologies	Technologies that automate and monitor power networks in real time to improve their efficiency, dependability, and sustainability.
Smart grids	Smart grids are electricity networks that use digital technology to monitor and manage energy flow, resulting in an efficient and reliable supply.
Smart Healthcare	The use of advanced technologies like IoT and AI in healthcare to improve diagnosis, treatment, and patient management.
Smart Transportation Systems	Smart transportation systems use technology and data to increase efficiency, safety, and sustainability.
Spam Filtering	Spam filtering is the use of algorithms and heuristics to detect and remove unsolicited and undesired email messages.
Spectral Efficiency	The rate of information sent over a certain bandwidth in a communication system, expressed in bits per second per Hertz.
Split Learning (SL)	A distributed learning method in which a model is divided into parts and trained on separate devices or servers, resulting in data privacy and efficiency.
Split Neural Networks (SplitNN)	A neural network training method in which the model is divided between the client and the server, with data processing dispersed between the two.
SplitFed Learning (SFL)	Combines federated and split learning, utilising both local and distributed model training while ensuring data privacy.

Static Reconfiguration Static reconfiguration is a one-time configuration of system characteristics or architecture that remains constant throughout operation.

Statistical Learning Statistical learning is the application of statistical methods and models to understand and predict data patterns.

Stereoscopic Projectors Stereoscopic projectors produce a 3D visual impression by showing two slightly distinct images to each eye.

Stochastic Gradient Descent (SGD) SGD is a variation of gradient descent in which the model is updated using a single or a few training instances at each iteration, hence increasing efficiency.

Supervised Learning Supervised Learning is a sort of machine learning in which the model is trained on labelled data before making predictions or classifying fresh data.

Support Vector Machine (SVM) A supervised learning technique for classification and regression applications that determines the best hyperplane to separate data points.

Text Categorization Text categorization is the process of grouping text into predetermined categories using machine learning and natural language processing algorithms.

Treatment Recommendations The application of data analysis and machine learning to recommend medical therapies based on specific patient circumstances and histories.

Unmanned Aerial Vehicle (UAV) networks UAV networks are networks of drones that communicate wirelessly to perform coordinated tasks such as surveillance or delivery.

Unmanned Aerial Vehicles (UAVs) These are aircraft systems without a human pilot on board, sometimes known as drones, that are employed for a variety of purposes such as surveillance and delivery.

Unmanned Aircraft Drones are employed for a variety of purposes, including surveillance, delivery, and precision agriculture.

URLLC Ultra-dependable Low-Latency Communications, a critical use case in 5G networks for applications that require highly dependable and time-sensitive communication.

Varactor-tuned Resonators Varactor-tuned resonators are those whose frequencies may be altered using varactor diodes, making them handy for tuning and filtering applications.

Virtual Assistants Siri and Alexa are examples of virtual assistants, which are AI-powered software agents that perform activities or services in response to user requests or queries.

Virtual Reality (VR) VR is a technology that provides a simulated environment for users to interact with, usually via head-mounted display.

Wearable Smart Gadgets	Wearable Smart Gadgets are wearable electronic gadgets that monitor and interact with the person and the environment.
Wireless Power Transfer (WPT)	It is the wireless transmission of electrical energy, often via electromagnetic fields, to power or charge devices from a remote location.
Wireless Sensor Network (WSN)	An interconnected system of geographically dispersed devices that record and track environmental conditions before relaying the results to a central location.

Index

www.ingramcontent.com/pod-product-compliance
Lightning Source LLC
LaVergne TN
LVHW011849200926
848484LV00002B/430
* 9 7 8 1 8 3 9 5 3 9 4 5 9 *